A Concise Introduction to Multiagent Systems and Distributed Artificial Intelligence

Synthesis Lectures on Artificial Intelligence and Machine Learning

Editors
Ronald J. Brachman, *Yahoo Research*

Tom Dietterich, *Oregon State University*

Intelligent Autonomous Robotics
Peter Stone
2007

A Concise Introduction to Multiagent Systems and Distributed Artificial Intelligence
Nikos Vlassis
2007

A Concise Introduction to Multiagent Systems and Distributed Artificial Intelligence

Nikos Vlassis

ISBN: 978-3-031-00415-5   paperback
ISBN: 978-3-031-00415-5   paperback

ISBN: 978-3-031-01543-4   ebook
ISBN: 978-3-031-01543-4   ebook

DOI: 10.1007/978-3-031-01543-4

A Publication in the Springer series
SYNTHESIS LECTURES ON ARTIFICIAL INTELLIGENCE AND MACHINE LEARNING
SEQUENCE IN SERIES: #2

Lecture #2
Series Editors: Ronald Brachman, Yahoo! Research and Thomas G. Dietterich, Oregon State University

First Edition
10 9 8 7 6 5 4 3 2 1

A Concise Introduction to Multiagent Systems and Distributed Artificial Intelligence

Nikos Vlassis
Department of Production Engineering and Management
Technical University of Crete
Greece

SYNTHESIS LECTURES ON ARTIFICIAL INTELLIGENCE AND MACHINE LEARNING SEQUENCE IN SERIES: #2

ABSTRACT

Multiagent systems is an expanding field that blends classical fields like game theory and decentralized control with modern fields like computer science and machine learning. This monograph provides a concise introduction to the subject, covering the theoretical foundations as well as more recent developments in a coherent and readable manner.

The text is centered on the concept of an agent as decision maker. Chapter 1 is a short introduction to the field of multiagent systems. Chapter 2 covers the basic theory of single-agent decision making under uncertainty. Chapter 3 is a brief introduction to game theory, explaining classical concepts like Nash equilibrium. Chapter 4 deals with the fundamental problem of coordinating a team of collaborative agents. Chapter 5 studies the problem of multiagent reasoning and decision making under partial observability. Chapter 6 focuses on the design of protocols that are stable against manipulations by self-interested agents. Chapter 7 provides a short introduction to the rapidly expanding field of multiagent reinforcement learning.

The material can be used for teaching a half-semester course on multiagent systems covering, roughly, one chapter per lecture.

Nikos Vlassis is Assistant Professor at the Department of Production Engineering and Management at the Technical University of Crete, Greece. His email is vlassis@dpem.tuc.gr

KEYWORDS

Multiagent Systems, Distributed Artificial Intelligence, Game Theory, Decision Making under Uncertainty, Coordination, Knowledge and Information, Mechanism Design, Reinforcement Learning.

Contents

Preface

This monograph is based on a graduate course on multiagent systems that I have taught at the University of Amsterdam, The Netherlands, from 2003 until 2006. This is the revised version of an originally unpublished manuscript that I wrote in 2003 and used as lecture notes. Since then the field has grown tremendously, and a large body of new literature has become available. Encouraged by the positive feedback I have received all these years from students and colleagues, I decided to compile this new, revised and up-to-date version.

Multiagent systems is a subject that has received much attention lately in science and engineering. It is a subject that blends classical fields like game theory and decentralized control with modern fields like computer science and machine learning. In the monograph I have tried to translate several of the concepts that appear in the above fields into a coherent and comprehensive framework for multiagent systems, aiming at keeping the text at a relatively introductory level without compromising its consistency or technical rigor. There is no mathematical prerequisite for the text; the covered material should be self-contained.

The text is centered on the concept of an agent as decision maker. The 1st chapter is an introductory chapter on multiagent systems. Chapter 2 addresses the problem of single-agent decision making, introducing the concepts of a Markov state and utility function. Chapter 3 is a brief introduction to game theory, in particular strategic games, describing classical solution concepts like iterated elimination of dominated actions and Nash equilibrium. Chapter 4 focuses on collaborative multiagent systems, and deals with the problem of multiagent coordination; it includes some standard coordination techniques like social conventions, roles, and coordination graphs. Chapter 5 examines the case where the perception of the agents is imperfect, and what consequences this may have in the reasoning and decision making of the agents; it deals with the concepts of information, knowledge, and common knowledge, and presents the model of a Bayesian game for multiagent decision making under partial observability. Chapter 6 deals with the problem of how to develop protocols that are nonmanipulable by a group of self-interested agents, discussing the revelation principle and the Vickrey-Clarke-Groves (VCG) mechanism. Finally, chapter 7 is a short introduction to reinforcement learning, that allows the agents to learn how to take good decisions; it covers the models of Markov decision processes and Markov games, and the problem of exploration.

The monograph can be used as teaching material in a half-semester course on multiagent systems; each chapter corresponds roughly to one lecture. This is how I have used the material in the past.

I am grateful to Jelle Kok, Frans Oliehoek, and Matthijs Spaan, for their valuable contributions and feedback. I am also thankful to Taylan Cemgil, Jan Nunnink, Dov Samet, Yoav Shoham, and Emilios Tigos, and numerous students at the University of Amsterdam for their comments on earlier versions of this manuscript. Finally I would like to thank Peter Stone for encouraging me to publish this work.

Nikos Vlassis
Chania, March 2007

CHAPTER 1

Introduction

In this chapter we give a brief introduction to multiagent systems, discuss their differences with single-agent systems, and outline possible applications and challenging issues for research.

1.1 MULTIAGENT SYSTEMS AND DISTRIBUTED AI

The modern approach to **artificial intelligence** (AI) is centered around the concept of a **rational agent**. An agent is anything that can perceive its environment through sensors and act upon that environment through actuators (Russell and Norvig, 2003). An agent that always tries to optimize an appropriate performance measure is called a 'rational agent'. Such a definition of a 'rational agent' is fairly general and can include human agents (having eyes as sensors, hands as actuators), robotic agents (having cameras as sensors, wheels as actuators), or software agents (having a graphical user interface as sensor and as actuator). From this perspective, AI can be regarded as the study of the principles and design of artificial rational agents.

However, agents are seldom stand-alone systems. In many situations they coexist and interact with other agents in several different ways. Examples include software agents on the Internet, soccer playing robots (see Fig. 1.1), and many more. Such a system that consists of a group of agents that can potentially interact with each other is called a **multiagent system** (MAS), and the corresponding subfield of AI that deals with principles and design of multiagent systems is called **distributed AI**.

1.2 CHARACTERISTICS OF MULTIAGENT SYSTEMS

What are the fundamental aspects that characterize a MAS and distinguish it from a single-agent system? One can think along the following dimensions.

1.2.1 Agent Design

It is often the case that the various agents that comprise a MAS are designed in different ways. The different design may involve the hardware, for example soccer robots based on different mechanical platforms, or the software, for example software agents (or 'softbots') running different code. Agents that are based on different hardware or implement different

FIGURE 1.1: A robot soccer team is an example of a multiagent system

behaviors are often called **heterogeneous**, in contrast to **homogeneous** agents that are designed in an identical way and have a priori the same capabilities. Agent heterogeneity can affect all functional aspects of an agent from perception to decision making.

1.2.2   Environment

Agents have to deal with environments that can be either **static** or **dynamic** (change with time). Most existing AI techniques for single agents have been developed for static environments because these are easier to handle and allow for a more rigorous mathematical treatment. In a MAS, the mere presence of multiple agents makes the environment appear dynamic from the point of view of each agent. This can often be problematic, for instance in the case of concurrently **learning** agents where non-stable behavior can be observed. There is also the issue of which parts of a dynamic environment an agent should treat as other agents and which not. We will discuss some of these issues in Chapter 7.

1.2.3   Perception

The collective information that reaches the sensors of the agents in a MAS is typically **distributed**: the agents may observe data that differ spatially (appear at different locations), temporally (arrive at different times), or semantically (require different interpretations). The fact that agents may observe different things makes the world **partially observable** to each agent, which has various consequences in the decision making of the agents. For instance, optimal multiagent planning under partial observability can be an intractable problem. An additional issue is **sensor fusion**, that is, how the agents can optimally combine their perceptions in order to increase their collective knowledge about the current state. In Chapter 5 we will discuss some of the above in more detail.

1.2.4   Control

Contrary to single-agent systems, the **control** in a MAS is typically decentralized. This means that the decision making of each agent lies to a large extent within the agent itself. Decentralized control is preferred over centralized control (that involves a center) for reasons of robustness and fault-tolerance. However, not all MAS protocols can be easily distributed, as we will see in Chapter 6. The general problem of multiagent decision making is the subject of **game theory** which we will briefly cover in Chapter 3. In a **collaborative** or **team** MAS where the agents share the same interests, distributed decision making offers asynchronous computation and speedups, but it also has the downside that appropriate **coordination** mechanisms need to be additionally developed. Chapter 4 is devoted to the topic of multiagent coordination.

1.2.5   Knowledge

In single-agent systems we typically assume that the agent knows its own actions but not necessarily how the world is affected by its actions. In a MAS, the levels of **knowledge** of each agent about the current world state can differ substantially. For example, in a team MAS involving two homogeneous agents, each agent may know the available action set of the other agent, both agents may know (by communication) their current perceptions, or they can infer the intentions of each other based on some shared prior knowledge. On the other hand, an agent that observes an adversarial team of agents will typically be unaware of their action sets and their current perceptions, and might also be unable to infer their plans. In general, in a MAS each agent must also consider the knowledge of each other agent in its decision making. In Chapter 5 we will discuss the concept of **common knowledge**, according to which every agent knows a fact, every agent knows that every other agent knows this fact, and so on.

1.2.6   Communication

Interaction is often associated with some form of **communication**. Typically we view communication in a MAS as a two-way process, where all agents can potentially be senders and receivers of messages. Communication can be used in several cases, for instance, for coordination among cooperative agents or for **negotiation** among **self-interested** agents. This additionally raises the issue of what network protocols to use in order for the exchanged information to arrive safely and timely, and what **language** the agents must speak in order to understand each other (especially, if they are heterogeneous). We will see throughout the book several examples of multiagent protocols involving communication.

1.3   APPLICATIONS

Just as with single-agent systems in traditional AI, it is difficult to anticipate the full range of applications where MASs can be used. Some applications have already appeared, for instance in software engineering where MAS technology has been recognized as a novel and promising software building paradigm: a complex software system can be treated as a collection of many small-size autonomous agents, each with its own local functionality and properties, and where interaction among agents enforces total system integrity. Some of the benefits of using MAS technology in large systems are (Sycara, 1998):

- Speedup and efficiency, due to the asynchronous and parallel computation.

- Robustness and reliability, in the sense that the whole system can undergo a 'graceful degradation' when one or more agents fail.

- Scalability and flexibility, since it is easy to add new agents to the system.

- Cost, assuming that an agent is a low-cost unit compared to the whole system.

- Development and reusability, since it is easier to develop and maintain a modular system than a monolithic one.

A very challenging application domain for MAS technology is the Internet. Today the Internet has developed into a highly distributed **open** system where heterogeneous software agents come and go, there are no well established protocols or languages on the 'agent level' (higher than TCP/IP), and the structure of the network itself keeps on changing. In such an environment, MAS technology can be used to develop agents that act on behalf of a user and are able to negotiate with other agents in order to achieve their goals. Electronic commerce and auctions are such examples (Cramton et al., 2006, Noriega and Sierra, 1999). One can also think of applications where agents can be used for distributed data mining and information retrieval (Kowalczyk and Vlassis, 2005, Symeonidis and Mitkas, 2006).

Other applications include sensor networks, where the challenge is to efficiently allocate resources and compute global quantities in a distributed fashion (Lesser et al., 2003, Paskin et al., 2005); social sciences, where MAS technology can be used for studying interactivity and other social phenomena (Conte and Dellarocas, 2001, Gilbert and Doran, 1994); robotics, where typical applications include distributed localization and decision making (Kok et al., 2005, Roumeliotis and Bekey, 2002); artificial life and computer games, where the challenge is to build agents that exhibit intelligent behavior (Adamatzky and Komosinski, 2005, Terzopoulos, 1999).

A recent popular application of MASs is robot soccer, where teams of real or simulated autonomous robots play soccer against each other (Kitano et al., 1997). Robot soccer provides a testbed where MAS algorithms can be tested, and where many real-world characteristics are present: the domain is continuous and dynamic, the behavior of the opponents may be difficult to predict, there is uncertainty in the sensor signals, etc. A related application is robot rescue, where teams of simulated or real robots must explore an unknown environment in order to discover victims, extinguish fires, etc. Both applications are organized by the RoboCup Federation (`www.robocup.org`).

1.4 CHALLENGING ISSUES

The transition from single-agent systems to MASs offers many potential advantages but also raises challenging issues. Some of these are:

- How to decompose a problem, allocate subtasks to agents, and synthesize partial results.

- How to handle the distributed perceptual information. How to enable agents to maintain consistent shared models of the world.

- How to implement decentralized control and build efficient coordination mechanisms among agents.

- How to design efficient multiagent planning and learning algorithms.

- How to represent knowledge. How to enable agents to reason about the actions, plans, and knowledge of other agents.

- How to enable agents to communicate. What communication languages and protocols to use. What, when, and with whom should an agent communicate.

- How to enable agents to negotiate and resolve conflicts.

- How to enable agents to form organizational structures like teams or coalitions. How to assign roles to agents.

- How to ensure coherent and stable system behavior.

Clearly the above problems are interdependent and their solutions may affect each other. For example, a distributed planning algorithm may require a particular coordination mechanism, learning can be guided by the organizational structure of the agents, and so on. In the later following chapters we will try to provide answers to some of the above questions.

1.5 NOTES AND FURTHER READING

The review articles of Sycara (1998) and Stone and Veloso (2000) provide concise and readable introductions to the field. The books of Huhns (1987), Singh (1994), O'Hare and Jennings (1996), Ferber (1999), Weiss (1999), Stone (2000), Yokoo (2000), Conte and Dellarocas (2001), Xiang (2002), Wooldridge (2002), Bordini et al. (2005), Vidal (2007), and Shoham and Leyton-Brown (2007) offer more extensive treatments, emphasizing different AI, societal, and computational aspects of multiagent systems.

CHAPTER 2

Rational Agents

In this chapter we describe what a rational agent is, we investigate some characteristics of an agent's environment like observability and the Markov property, and we examine what is needed for an agent to behave optimally in an uncertain world where actions do not always have the desired effects.

2.1 WHAT IS AN AGENT?

Following Russell and Norvig (2003), an **agent** is anything that can be viewed as perceiving its **environment** through sensors and acting upon that environment through actuators.[1] Examples include humans, robots, or software agents. We often use the term **autonomous** to refer to an agent whose decision making relies to a larger extent on its own perception than to prior knowledge given to it at design time.

In this chapter we will study the problem of **optimal decision making** of an agent. That is, how an agent can choose the best possible action at each time step, given what it knows about the world around it. We will say that an agent is **rational** if it always selects an action that optimizes an appropriate **performance measure**, given what the agent knows so far. The performance measure is typically defined by the user (the designer of the agent) and reflects what the user expects from the agent in the task at hand. For example, a soccer robot must act so as to maximize the chance of scoring for its team, a software agent in an electronic auction must try to minimize expenses for its designer, and so on. A rational agent is also called an **intelligent** agent.

In the following we will mainly focus on **computational** agents, that is, agents that are explicitly designed for solving a particular task and are implemented on some computing device.

2.2 AGENTS AS RATIONAL DECISION MAKERS

The problem of decision making of an agent is a subject of **optimal control** (Bellman, 1961, Bertsekas, 2001). For the purpose of our discussion we will assume a discrete set of time steps $t = 0, 1, 2, \ldots$, in each of which the agent must choose an action a_t from a finite set of

---

[1]In this chapter we will use 'it' to refer to an agent, to emphasize that we are talking about computational entities.

actions A that it has available. Intuitively, in order to act rationally at time t, an agent should take both the past and the future into account when choosing an action. The past refers to what the agent has perceived and what actions it has taken until time t, and the future refers to what the agent expects to perceive and do after time t.

If we denote by θ_τ the observation of an agent at time τ, then the above implies that in order for an agent to optimally choose an action at time t, it must in general use its complete **history** of observations θ_τ and actions a_τ for $\tau \leq t$. The function

$$\pi(\theta_0, a_0, \theta_1, a_1, \ldots, \theta_t) = a_t \qquad (2.1)$$

that in principle would require mapping the complete history of observation–action pairs up to time t to an optimal action a_t, is called the **policy** of the agent.

As long as we can find a function π that implements the above mapping, the part of optimal decision making that refers to the past is solved. However, defining and implementing such a function is problematic; the complete history can consist of a very large (even infinite) number of observation–action pairs, which can vary from one task to another. Merely storing all observations would require very large memory, aside from the computational cost for actually computing π.

This fact calls for simpler policies. One possibility is for the agent to ignore all its percept history except for the last observation θ_t. In this case its policy takes the form

$$\pi(\theta_t) = a_t \qquad (2.2)$$

which is a mapping from the current observation of the agent to an action. An agent that simply maps its current observation θ_t to a new action a_t, thus effectively ignoring the past, is called a **reflex** agent, and its policy (2.2) is called **reactive** or **memoryless**. A natural question to ask is how successful a reflex agent can be. As we will see next, for a particular class of environments a reflex agent can do pretty well.

2.3   OBSERVABLE WORLDS AND THE MARKOV PROPERTY

From the discussion above it is clear that the terms 'agent' and 'environment' are coupled, so that one cannot be defined without the other Sutton and Barto (1998, ch. 3) discuss this point). For our purposes we will assume hereafter the existence of a **world** in which one or more agents are embedded, and in which they perceive, think, and act. The collective information that is contained in the world at any time step t, and that is relevant for the task at hand, will be called a **state** of the world and denoted by s_t. The set of all states of the world will be denoted by S. As an example, in a robot soccer game a world state can be characterized by the soccer field layout, the positions and velocities of all players and the ball, what each agent knows about each other, and other parameters that are relevant to the decision making of the agents like the elapsed time since the game started, etc.

Depending on the nature of the problem, a world can be either **discrete** or **continuous**. A discrete world can be characterized by a finite number of states, like the possible board configurations in a chess game. A continuous world can have infinitely many states, like the possible configurations of a point robot that translates freely on the plane in which case $S = \mathbb{R}^2$. Most of the existing AI techniques have been developed for discrete worlds, and this will be our main focus as well.

2.3.1 Observability

A fundamental property that characterizes a world from the point of view of an agent is related to the perception of the agent. We will say that the world is **(fully) observable** to an agent if the current observation θ_t of the agent completely reveals the current state of the world, that is, $s_t = \theta_t$. On the other hand, in a **partially observable** world the current observation θ_t of the agent provides only partial information about the current state s_t in the form of a deterministic or stochastic **observation model**, for instance a conditional probability distribution $p(s_t|\theta_t)$. The latter would imply that the current observation θ_t does not fully reveal the true world state, but to each state s_t the agent assigns probability $p(s_t|\theta_t)$ that s_t is the true state (with $0 \le p(s_t|\theta_t) \le 1$ and $\sum_{s_t \in S} p(s_t|\theta_t) = 1$). Here we treat s_t as a random variable that can take all possible values in S. The stochastic coupling between s_t and θ_t may alternatively be defined by an observation model in the form $p(\theta_t|s_t)$, and a *posterior* state distribution $p(s_t|\theta_t)$ can be computed from a *prior* distribution $p(s_t)$ using the **Bayes rule**:

$$p(s_t|\theta_t) = \frac{p(\theta_t|s_t)p(s_t)}{p(\theta_t)}. \tag{2.3}$$

Partial observability can in principle be attributed to two factors. First, it can be the result of **noise** in the agent's sensors. For example, due to sensor malfunction, the same state may 'generate' different observations to the agent at different points in time. That is, every time the agent visits a particular state it may perceive something different. Second, partial observability can be related to an inherent property of the environment referred to as **perceptual aliasing**: different states may produce identical observations to the agent at different time steps. In other words, two states may 'look' the same to an agent, although the states are different from each other. For example, two identical doors along a corridor will look exactly the same to the eyes of a human or the camera of a mobile robot, no matter how accurate each sensor system is.

Partial observability is much harder to handle than full observability, and algorithms for optimal decision making in a partially observable world can often become intractable. As we will see in Chapter 5, partial observability may affect not only what each agent knows about the world state, but also what each agent knows about each other's knowledge.

2.3.2   The Markov Property

Let us consider again the case of a reflex agent with a reactive policy $\pi(\theta_t) = a_t$ in a fully observable world. The assumption of observability implies $s_t = \theta_t$, and therefore the policy of the agent reads

$$\pi(s_t) = a_t. \qquad (2.4)$$

In other words, in an observable world the policy of a reflex agent is a mapping from world states to actions. The gain comes from the fact that in many problems the state of the world at time t provides a *complete characterization* of the history before time t. Such a world state that summarizes all relevant information about the past is said to be **Markov** or to have the **Markov property**. As we conclude from the above, in a Markov world an agent can safely use the memoryless policy (2.4) for its decision making, in place of the memory-expensive policy (2.1).

So far we have discussed how the policy of an agent may depend on its past experience and the particular characteristics of the environment. However, as we argued at the beginning, optimal decision making should also take the future into account. This is what we are going to examine next.

2.4   STOCHASTIC TRANSITIONS AND UTILITIES

As mentioned above, at each time step t the agent chooses an action a_t from a finite set of actions A. When the agent takes an action, the world changes as a result of this action. A **transition model** (or **world model**) specifies how the world changes when an action is executed. If the current world state is s_t and the agent takes action a_t, we can distinguish the following two cases:

- In a **deterministic** world, the transition model maps a state–action pair (s_t, a_t) to a single new state s_{t+1}. In chess, for example, every move changes the configuration on the board in a deterministic manner.

- In a **stochastic** world, the transition model maps a state–action pair (s_t, a_t) to a probability distribution $p(s_{t+1}|s_t, a_t)$ over states. As in the partial observability case above, s_{t+1} is a random variable that can take all possible values in S, each with corresponding probability $p(s_{t+1}|s_t, a_t)$. Most real-world applications involve stochastic transition models; for example, robot motion is inaccurate because of wheel slip and other effects.

We saw in the previous section that sometimes partial observability can be attributed to uncertainty in the perception of the agent. Here we see another example where uncertainty plays a role; namely, in the way the world changes when the agent executes an action. In a stochastic world, the effects of the actions of the agent are not known a priori. Instead, there is a random element that decides how the world changes as a result of an action. Clearly, stochasticity in the state transitions introduces an additional difficulty in the optimal decision making task of the agent.

2.4.1 From Goals to Utilities

In classical AI, a **goal** for a particular task is a desired state of the world. Accordingly, **planning** is defined as a search through the state space for an optimal path to the goal. When the world is deterministic, planning comes down to a graph search problem for which a variety of methods exist (Russell and Norvig, 2003, ch. 3).

In a stochastic world, however, planning cannot be done by simple graph search because transitions between states are nondeterministic. The agent must now take the uncertainty of the transitions into account when planning. To see how this can be realized, note that in a deterministic world an agent prefers by default a goal state to a non-goal state. More generally, an agent may hold **preferences** between any world states. For example, a soccer agent will mostly prefer to score a goal, will prefer less (but still a lot) to stand with the ball in front of an empty goal, and so on.

A way to formalize the notion of state preferences is by assigning to each state s a real number $U(s)$ that is called the **utility** of state s for that particular agent. Formally, for two states s and s' holds $U(s) > U(s')$ if and only if the agent prefers state s to state s', and $U(s) = U(s')$ if and only if the agent is indifferent between s and s'. Intuitively, the utility of a state expresses the 'desirability' of that state for the particular agent; the larger the utility of the state, the better the state is for that agent. In the discrete world of Fig. 2.1, for instance, an agent would prefer state d3 than state b2 or d2. Note that in a multiagent system, a state may be desirable to a particular agent and at the same time be undesirable to an another agent; in soccer, for example, scoring is typically unpleasant to the opponent agents.

4				
3				+1
2		−1		−1
1	start			
	a	b	c	d

FIGURE 2.1: A world with one desired (+1) and two undesired (−1) states

2.4.2   Decision Making in a Stochastic World

Equipped with utilities, the question now is how an agent can effectively use them for its decision making. Let us assume that there is only one agent in the world, and the world is stochastic with transition model $p(s_{t+1}|s_t, a_t)$. Suppose that the current state is s_t, and the agent is pondering how to choose its action a_t. Let $U(s)$ be the utility function for the particular agent. Utility-based decision making is based on the premise that the optimal action a_t^* of the agent at state s_t should maximize **expected utility**, that is,

$$a_t^* = \arg\max_{a_t \in A} \sum_{s_{t+1}} p(s_{t+1}|s_t, a_t)U(s_{t+1}) \tag{2.5}$$

where we sum over all possible states $s_{t+1} \in S$ the world may transition to, given that the current state is s_t and the agent takes action a_t. In words, to see how good an action is, the agent has to multiply the utility of each possible resulting state with the probability of actually reaching this state, and sum up over all states. Then the agent must choose the action a_t^* that gives the highest sum.

If each world state possesses a utility value, the agent can do the above calculations and compute an optimal action for each possible state. This provides the agent with a policy that maps states to actions in an optimal sense (optimal with respect to the given utilities). In particular, given a set of optimal (that is, highest attainable) utilities $U^*(s)$ in a given task, the **greedy** policy

$$\pi^*(s) = \arg\max_a \sum_{s'} p(s'|s, a)U^*(s') \tag{2.6}$$

is an **optimal policy** for the agent.

There is an alternative and often useful way to characterize an optimal policy. For each state s and each possible action a we can define an optimal **action value** or **Q-value** $Q^*(s, a)$ that measures the 'goodness' of action a in state s for that agent. For the Q-values holds $U^*(s) = \max_a Q^*(s, a)$, while an optimal policy can be computed as

$$\pi^*(s) = \arg\max_a Q^*(s, a) \tag{2.7}$$

which is a simpler formula than (2.6) that does not make use of a transition model. In Chapter 7 we will see how we can compute optimal Q-values $Q^*(s, a)$, and hence an optimal policy, in a given task.

2.4.3   Example: A Toy World

Let us close the chapter with an example, similar to the one used by Russell and Norvig (2003, ch. 21). Consider the world of Fig. 2.1 where in any state the agent can choose any one of

4	0.818 ($\rightarrow$)	0.865 ($\rightarrow$)	0.911 ($\rightarrow$)	0.953 ($\downarrow$)
3	0.782 ($\uparrow$)	0.827 ($\uparrow$)	0.907 ($\rightarrow$)	+1
2	0.547 ($\uparrow$)	−1	0.492 ($\uparrow$)	−1
1	0.480 ($\uparrow$)	0.279 ($\leftarrow$)	0.410 ($\uparrow$)	0.216 ($\leftarrow$)
	a	b	c	d

FIGURE 2.2: Optimal utilities and an optimal policy of the agent

the actions {*Up, Down, Left, Right*}. We assume that the world is fully observable (the agent always knows where it is), and stochastic in the following sense: every action of the agent to an intended direction succeeds with probability 0.8, but with probability 0.2 the agent ends up perpendicularly to the intended direction. Bumping on the border leaves the position of the agent unchanged. There are three terminal states, a desired one (the 'goal' state) with utility +1, and two undesired ones with utility −1. The initial position of the agent is a1.

We stress again that although the agent can perceive its own position and thus the state of the world, it cannot predict the effects of its actions on the world. For example, if the agent is in state c2, it knows that it is in state c2. However, if it tries to move *Up* to state c3, it may reach the intended state c3 (this will happen in 80% of the cases) but it may also reach state b2 (in 10% of the cases) or state d2 (in the rest 10% of the cases).

Assume now that optimal utilities have been computed for all states, as shown in Fig. 2.2. Applying the principle of maximum expected utility, the agent computes that, for instance, in state b3 the optimal action is *Up*. Note that this is the only action that avoids an accidental transition to state b2. Similarly, by using (2.6) the agent can now compute an optimal action for every state, which gives the optimal policy shown in parentheses.

Note that, unlike path planning in a deterministic world that can be described as graph search, decision making in stochastic domains requires computing a complete policy that maps states to actions. Again, this is a consequence of the fact that the results of the actions of an agent are unpredictable. Only after the agent has executed its action it can observe the new state of the world, from which it can select another action based on its precomputed policy.

2.5   NOTES AND FURTHER READING

We have mainly followed Chapters 2, 16, and 17 of the book of Russell and Norvig (2003) which we strongly recommend for further reading. An illuminating discussion on the agent–environment interface and the Markov property can be found in Chapter 3 of the book of Sutton and Barto (1998) which is another excellent text on agents and decision making. Bertsekas (2001) provides a more technical exposition. Spaan and Vlassis (2005) outline recent advances in the topic of sequential decision making under partial observability.

CHAPTER 3

Strategic Games

In this chapter we study the problem of **multiagent decision making** where a group of agents coexist in an environment and take simultaneous decisions. We use game theory to analyze the problem. In particular, we describe the model of a strategic game and we examine two fundamental solution concepts, iterated elimination of strictly dominated actions and Nash equilibrium.

3.1 GAME THEORY

As we saw in Chapter 2, an agent will typically be uncertain about the effects of its actions to the environment, and it has to take this uncertainty into account in its decision making. In a multiagent system where many agents take decisions at the same time, an agent will also be uncertain about the decisions of the other participating agents. Clearly, what an agent should do depends on what the other agents will do.

Multiagent decision making is the subject of **game theory** (Osborne and Rubinstein, 1994). Although originally designed for modeling economical interactions, game theory has developed into an independent field with solid mathematical foundations and many applications. The theory tries to understand the behavior of interacting agents under conditions of uncertainty, and is based on two premises. First, that the participating agents are **rational**. Second, that they reason **strategically**, that is, they take into account the other agents' decisions in their decision making.

Depending on the way the agents choose their actions, there are different types of games. In a **strategic game** each agent chooses his[1] strategy only once at the beginning of the game, and then all agents take their actions simultaneously. In an **extensive game** the agents are allowed to reconsider their plans during the game, and they may be imperfectly informed about the actions played by the other agents. In this chapter we will only consider strategic games.

---

[1]In this chapter we will use 'he' or 'she' to refer to an agent, following the convention in the literature (Osborne and Rubinstein, 1994, p. xiii).

3.2   STRATEGIC GAMES

A strategic game, or game in **normal form**, is the simplest game-theoretic model of agents'
interaction. It can be viewed as a multiagent extension of the decision-theoretic model of Chapter 2, and is characterized by the following elements:

- There are $n > 1$ agents in the world.

- Each agent i can choose an action, or **strategy**, a_i from his own action set A_i. The
tuple $(a_1, \ldots, a_n)$ of individual actions is called a **joint action** or an **action profile**,
and is denoted by a or (a_i). We will use the notation a_{-i} to refer the actions of all
agents except i, and (a_i, a_{-i}) or $[a_i, a_{-i}]$ to refer to a joint action where agent i takes a
particular action a_i.

- The game is 'played' on a fixed world state s (we are not concerned with state transitions
here). The state can be defined as consisting of the n agents, their action sets A_i, and
their payoffs, as we explain next.

- Each agent i has his own action value function $Q_i^*(s, a)$ that measures the goodness of
the *joint action a for the agent i*. Note that each agent may assign different preferences
to different joint actions. Since s is fixed, we drop the symbol s and instead use
$u_i(a) \equiv Q_i^*(s, a)$, which is called the **payoff function** of agent i. We assume that the
payoff functions are predefined and fixed. (We will deal with the case of learning the
payoff functions in Chapter 7.)

- The state is fully observable to all agents. That is, all agents know (i) each other, (ii) the
action sets of each other, and (iii) the payoffs of each other. More strictly, the primitives
(i)-(iii) of the game are **common knowledge** among agents. That is, all agents know
(i)–(iii), they all know that they all know (i)–(iii), and so on to any depth. (We will
discuss common knowledge in detail in Chapter 5.)

- Each agent chooses a single action; it is a single-shot game. Moreover, all agents choose
their actions simultaneously and independently; no agent is informed of the decision
of any other agent prior to making his own decision.

In summary, in a strategic game each agent chooses a single action, and then he receives
a payoff that depends on the selected joint action. This joint action is called the **outcome** of
the game. Although the payoff functions of the agents are common knowledge, an agent does
not know in advance the action choices of the other agents. The best he can do is to try to
predict the actions of the other agents. A **solution** to a game is a prediction of the outcome of
the game using the assumption that all agents are rational and strategic.

	Not confess	Confess
Not confess	3, 3	0, 4
Confess	4, 0	1, 1

FIGURE 3.1: The prisoner's dilemma

In the special case of two agents, a strategic game can be graphically represented by a **payoff matrix,** where the rows correspond to the actions of agent 1, the columns to the actions of agent 2, and each entry of the matrix contains the payoffs of the two agents for the corresponding joint action. In Fig. 3.1 we show the payoff matrix of a classical game, the **prisoner's dilemma,** whose story goes as follows:

Two suspects in a crime are independently interrogated. If they both confess, each will spend three years in prison. If only one confesses, he will run free while the other will spend four years in prison. If neither confesses, each will spend one year in prison.

In this example each agent has two available actions, *Not confess* or *Confess.* Translating the above story into appropriate payoffs for the agents, we get in each entry of the matrix the pairs of numbers that are shown in Fig. 3.1 (note that a payoff is by definition a 'reward', whereas spending three years in prison is a 'penalty'). For example, the entry (4, 0) indicates that if the first agent confesses and the second agent does not, then the first agent will get payoff 4 and the second agent will get payoff 0.

In Fig. 3.2 we see two more examples of strategic games. The game in Fig. 3.2(a) is known as 'matching pennies'; each of two agents chooses either *Head* or *Tail.* If the choices differ, agent 1 pays agent 2 a cent; if they are the same, agent 2 pays agent 1 a cent. Such a game is called **strictly competitive** or **zero-sum** because $u_1(a) + u_2(a) = 0$ for all a. The game in Fig. 3.2(b) is played between two car drivers at a crossroad; each agent wants to cross first (and he will get payoff 1), but if they both cross they will crash (and get payoff -1). Such a game is called a **coordination game** (we will study coordination games in Chapter 4).

What does game theory predict that a rational agent will do in the above examples? In the next sections we will describe two fundamental solution concepts for strategic games.

	Head	Tail			Cross	Stop
Head	1, −1	−1, 1		Cross	−1, 1	1, 0
Tail	−1, 1	1, −1		Stop	0, 1	0, 0

(a)   (b)

FIGURE 3.2: A strictly competitive game (a), and a coordination game (b)

3.3 ITERATED ELIMINATION OF DOMINATED ACTIONS

The first solution concept is based on the assumption that a rational agent will never choose a suboptimal action. With suboptimal we mean an action that, no matter what the other agents do, will always result in lower payoff for the agent than some other action. We formalize this as follows:

Definition 3.1. *We will say that an action a_i of agent i is* **strictly dominated** *by another action a_i' of agent i if*

$$u_i(a_i', a_{-i}) > u_i(a_i, a_{-i}) \qquad (3.1)$$

for all actions a_{-i} of the other agents.

In the above definition, $u_i(a_i, a_{-i})$ is the payoff the agent i receives if he takes action a_i while the other agents take a_{-i}. In the prisoner's dilemma, for example, *Not confess* is a strictly dominated action for agent 1; no matter what agent 2 does, the action *Confess* always gives agent 1 higher payoff than the action *Not confess* (4 as opposed to 3 if agent 2 does not confess, and 1 as opposed to 0 if agent 2 confesses). Similarly, *Not confess* is a strictly dominated action for agent 2.

Iterated elimination of strictly dominated actions (IESDA) is a solution technique that iteratively eliminates strictly dominated actions from all agents, until no more actions are strictly dominated. It is solely based on the following two assumptions:

• A rational agent would never take a strictly dominated action.

• It is common knowledge that all agents are rational.

As an example, we will apply IESDA to the prisoner's dilemma. As we explained above, the action *Not confess* is strictly dominated by the action *Confess* for both agents. Let us start from agent 1 by eliminating the action *Not confess* from his action set. Then the game reduces to a single-row payoff matrix where the action of agent 1 is fixed (*Confess*) and agent 2 can choose between *Not confess* and *Confess*. Since the latter gives higher payoff to agent 2 (4 as opposed to 3), agent 2 will prefer *Confess* to *Not confess*. Thus IESDA predicts that the outcome of the prisoner's dilemma will be (*Confess, Confess*).

As another example consider the game of Fig. 3.3(a) where agent 1 has two actions U and D and agent 2 has three actions L, M, and R. It is easy to verify that in this game IESDA will predict the outcome (U, M) by first eliminating R (strictly dominated by M), then D, and finally L. However, IESDA may sometimes produce very inaccurate predictions for a game, as in the two games of Fig. 3.2 and also in the game of Fig. 3.3(b) where no actions can be eliminated. In these games IESDA essentially predicts that *any* outcome is possible.

	L	M	R
U	1,0	1,2	0,1
D	0,3	0,1	2,0

(a)

	L	M	R
U	1,0	1,2	0,1
D	0,3	0,1	2,2

(b)

FIGURE 3.3: Examples where IESDA predicts a single outcome (a), or predicts that any outcome is possible (b).

A characteristic of IESDA is that the agents do not need to maintain *beliefs* about the other agents' strategies in order to compute their optimal actions. The only thing that is required is the common knowledge assumption that each agent is rational. Moreover, it can be shown that the algorithm is insensitive to the speed and the elimination order; it will always produce the same result no matter how many actions are eliminated in each step and in which order. However, as we saw in the examples above, IESDA can sometimes fail to make useful predictions for the outcome of a game.

3.4   NASH EQUILIBRIUM

A **Nash equilibrium** (NE) is a stronger solution concept than IESDA, in the sense that it produces more accurate predictions in a wider class of games. It can be formally defined as follows:

Definition 3.2. *A Nash equilibrium is a joint action a* with the property that for every agent i holds*

$$u_i(a_i^*, a_{-i}^*) \geq u_i(a_i, a_{-i}^*) \tag{3.2}$$

for all actions $a_i \in A_i$.

In other words, a NE is a joint action from where no agent can *unilaterally* improve his payoff, and therefore no agent has any incentive to deviate. Note that, contrary to IESDA that describes a solution of a game by means of an algorithm, a NE describes a solution in terms of the conditions that hold at that solution.

There is an alternative definition of a NE that makes use of the so-called **best-response** function. For agent i, this function is defined as

$$B_i(a_{-i}) = \{a_i \in A_i : u_i(a_i, a_{-i}) \geq u_i(a_i', a_{-i}) \text{ for all } a_i' \in A_i\}, \tag{3.3}$$

and $B_i(a_{-i})$ can be a set containing many actions. In the prisoner's dilemma, for example, when agent 2 takes the action *Not confess*, the best-response of agent 1 is the action *Confess* (because 4 > 3). Similarly, we can compute the best-response function of each agent:

$$B_1(\textit{Not confess}) \quad = \quad \textit{Confess},$$
$$B_1(\textit{Confess}) \quad\quad = \quad \textit{Confess},$$
$$B_2(\textit{Not confess}) \quad = \quad \textit{Confess},$$
$$B_2(\textit{Confess}) \quad\quad = \quad \textit{Confess}.$$

In this case, the best-response functions are singleton-valued. Using the definition of a best-response function we can now formulate the following:

Definition 3.3. *A Nash equilibrium is a joint action a^* with the property that for every agent i holds*

$$a_i^* \in B_i(a_{-i}^*). \tag{3.4}$$

That is, at a NE, each agent's action is an optimal response to the other agents' actions. In the prisoner's dilemma, for instance, given that $B_1(\textit{Confess}) = \textit{Confess}$, and $B_2(\textit{Confess}) = \textit{Confess}$, we conclude that $(\textit{Confess}, \textit{Confess})$ is a NE. Moreover, we can easily show the following:

Proposition 3.1. The two definitions 3.2 and 3.3 of a NE are equivalent.

Proof. Suppose that (3.4) holds. Then, using (3.3) we see that for each agent i, the action a_i^* must satisfy $u_i(a_i^*, a_{-i}^*) \geq u_i(a_i', a_{-i}^*)$ for all $a_i' \in A_i$. The latter is precisely the definition of a NE according to (3.2). Similarly for the converse. □

The definitions 3.2 and 3.3 suggest a brute-force method for finding the Nash equilibria of a game: enumerate all possible joint actions and then verify which ones satisfy (3.2) or (3.4). Note that the cost of such an algorithm is exponential in the number of agents.

It turns out that a strategic game can have zero, one, or more than one Nash equilibria. For example, (*Confess*, *Confess*) is the only NE in the prisoner's dilemma. We also find that the zero-sum game in Fig. 3.2(a) does not have a NE, while the coordination game in Fig. 3.2(b) has two Nash equilibria (*Cross*, *Stop*) and (*Stop*, *Cross*). Similarly, (*U*, *M*) is the only NE in both games of Fig. 3.3.

We argued above that a NE is a stronger solution concept than IESDA in the sense that it produces more accurate predictions of a game. For instance, the game of Fig. 3.3(b) has only one NE, but IESDA predicts that any outcome is possible. In general, we can show the following two propositions (the proof of the second is left as an exercise):

Proposition 3.2. A NE always survives IESDA.

Proof. Let a^* be a NE, and let us assume that a^* does not survive IESDA. This means that for some agent i the component a_i^* of the action profile a^* is strictly dominated by another action a_i of agent i. But then (3.1) implies that $u_i(a_i, a_{-i}^*) > u_i(a_i^*, a_{-i}^*)$ which contradicts the Definition 3.2 of a NE. □

Proposition 3.3. If IESDA eliminates all but a single joint action a, then a is the unique NE of the game.

Note also that in the prisoner's dilemma, the joint action (*Not confess*, *Not confess*) gives both agents payoff 3, and thus it should have been the preferable choice. However, from this joint action each agent has an incentive to deviate, to be a 'free rider'. Only if the agents had made an agreement in advance, and only if trust between them was common knowledge, would they have opted for this non-equilibrium joint action which is optimal in the following sense:

Definition 3.4. *A joint action a is* **Pareto optimal** *if there is no other joint action a' for which $u_i(a') \geq u_i(a)$ for each i and $u_j(a') > u_j(a)$ for some j.*

So far we have implicitly assumed that when the game is actually played, each agent i will choose his action deterministically from his action set A_i. This is however not always true. In many cases there are good reasons for an agent to introduce randomness in his behavior; for instance, to avoid being predictable when he repeatedly plays a zero-sum game. In these cases an agent i can choose actions a_i according to some probability distribution:

Definition 3.5. *A* **mixed strategy** *for an agent i is a probability distribution over his actions $a_i \in A_i$.*

In his celebrated theorem, Nash (1950) showed that a strategic game with a finite number of agents and a finite number of actions always has an equilibrium in mixed strategies. Osborne and Rubinstein (1994, sec. 3.2) give several interpretations of such a mixed strategy Nash equilibrium. Porter et al. (2004) and von Stengel (2007) describe several algorithms for computing Nash equilibria, a problem whose complexity has been a long-standing issue (Papadimitriou, 2001).

3.5   NOTES AND FURTHER READING

The book of von Neumann and Morgenstern (1944) and the half-page long article of Nash (1950) are classics in game theory. The book of Osborne and Rubinstein (1994) is the standard textbook on game theory, and it is highly recommended. The book of Gibbons (1992) and the book of Osborne (2003) offer a readable introduction to the field, with several applications. Russell and Norvig (2003, ch. 17) also include an introductory section on game theory. The book of Nisan et al. (2007) focuses on computational aspects of game theory.

CHAPTER 4

Coordination

In this chapter we address the problem of multiagent **coordination**. We analyze the problem using the framework of strategic games that we studied in Chapter 3, and we describe several practical techniques like social conventions, roles, and coordination graphs.

4.1 COORDINATION GAMES

As we argued in Chapter 1, decision making in a multiagent system should preferably be carried out in a decentralized manner for reasons of efficiency and robustness. This additionally requires developing a coordination mechanism. In the case of **collaborative** agents, coordination ensures that the agents do not obstruct each other when taking actions, and that these actions serve the common goal of the team (for example, two teammate soccer robots must coordinate their actions when deciding who should go for the ball). Informally, coordination can be regarded as the process by which the individual decisions of the agents result in good joint decisions for the group.

Formally, we can model a coordination problem as a **coordination game** using the tools of game theory, and solve it according to some solution concept, for instance Nash equilibrium. We have already seen an example in Fig. 3.2(b) of Chapter 3 of a strategic game where two cars meet at a crossroad and one driver should cross and the other one should stop. That game has two Nash equilibria, (*Cross*, *Stop*) and (*Stop*, *Cross*). In the case of n collaborative agents, all agents in the team share the same payoff function $u_1(a) = \ldots = u_n(a) \equiv u(a)$ in the corresponding coordination game. Figure 4.1 shows an example of a coordination game (played between two agents who want to go to the movies together) that also has two Nash equilibria. Generalizing from these two examples, we can formally define coordination as *the process in which a group of agents choose a single Pareto optimal Nash equilibrium in a game.*

	Thriller	Comedy
Thriller	1, 1	0, 0
Comedy	0, 0	1, 1

FIGURE 4.1: A coordination game

In Chapter 3 we described a Nash equilibrium in terms of the conditions that hold at the equilibrium point, and disregarded the issue of how the agents can actually reach this point. Coordination is a more earthy concept, as it asks how the agents can actually agree on a single equilibrium in a game that involves more than one equilibria. Reducing coordination to the problem of equilibrium selection in a game allows for the application of existing techniques from game theory (Harsanyi and Selten, 1988). In the rest of this chapter we will focus on some simple coordination techniques that can be readily implemented in practical systems. We will throughout assume that the agents are collaborative (they share the same payoff function), and that they have perfect information about the game primitives (see Section 3.2). Also by 'equilibrium' we will mean here 'Pareto optimal Nash equilibrium', unless otherwise stated.

4.2   SOCIAL CONVENTIONS

As we saw above, in order to solve a coordination problem, a group of agents are faced with the problem of how to choose their actions in order to select the same equilibrium in a game. Clearly, there can be no recipe to tell the agents which equilibrium to choose in every possible game they may play in the future. Nevertheless, we can devise recipes that will instruct the agents on how to choose a single equilibrium in any game. Such a recipe will be able to guide the agents in their action selection procedure.

A **social convention** (or **social law**) is such a recipe that places constraints on the possible action choices of the agents. It can be regarded as a rule that dictates how the agents should choose their actions in a coordination game in order to reach an equilibrium. Moreover, given that the convention has been established and is common knowledge among agents, no agent can benefit from not abiding by it.

Boutilier (1996) has proposed a general convention that achieves coordination in a large class of systems and is very easy to implement. The convention assumes a unique ordering scheme of joint actions that is common knowledge among agents. In a particular game, each agent first computes all equilibria of the game, and then selects the first equilibrium according to this ordering scheme. For instance, a lexicographic ordering scheme can be used in which the agents are ordered first, and then the actions of each agent are ordered. In the coordination game of Fig. 4.1, for example, we can order the agents lexicographically by $1 \succ 2$ (meaning that agent 1 has 'priority' over agent 2), and the actions by $Thriller \succ Comedy$. The first equilibrium in the resulting ordering of joint actions is (*Thriller*, *Thriller*) and this will be the unanimous choice of the agents. Given that a single equilibrium has been selected, each agent can then choose his individual action as the corresponding component of the selected equilibrium.

When the agents can perceive more aspects of the world state than just the primitives of the game (actions and payoffs), one can think of more elaborate ordering schemes for coordination. Consider the traffic game of Fig. 3.2(b), for example, as it is 'played' in the real world. Besides the game primitives, the state now also contains the relative orientation of the cars in the physical environment. If the state is fully observable by both agents (and this fact is common knowledge), then a simple convention is that the driver coming from the right will always have priority over the other driver in the lexicographic ordering. If we also order the actions by *Cross* ≻ *Stop*, then coordination by social conventions implies that the driver from the right will cross the road first. Similarly, if traffic lights are available, the established convention is that the driver who sees the red light must stop.

When communication is available, we only need to impose an ordering $i = 1, \ldots, n$ of the agents that is common knowledge. Coordination can now be achieved by the following algorithm: Each agent i (except agent 1) waits until all previous agents $1, \ldots, i-1$ in the ordering have broadcast their chosen actions, and then agent i computes its component a_i^* of an equilibrium that is consistent with the choices of the previous agents and broadcasts a_i^* to all agents that have not chosen an action yet. Note that here the fixed ordering of the agents together with the wait/send primitives result in a synchronized sequential execution order of the coordination algorithm.

4.3   ROLES

Coordination by social conventions relies on the assumption that an agent can compute all equilibria in a game before choosing a single one. However, computing equilibria can be expensive when the action sets of the agents are large, so it makes sense to try to reduce the size of the action sets first. Such a reduction can have computational advantages in terms of speed, but it can also simplify the equilibrium selection problem; in some cases the resulting subgame contains only one equilibrium which is trivial to find.

A natural way to reduce the action sets of the agents is by assigning **roles** to the agents. Formally, a role can be regarded as a masking operator on the action set of an agent given a particular state. In practical terms, if an agent is assigned a role at a particular state, then some of the agent's actions are deactivated at this state. In soccer for example, an agent that is currently in the role of defender cannot attempt to *Score*.

A role can facilitate the solution of a coordination game by reducing it to a subgame where the equilibria are easier to find. For example, in Fig. 4.1, if agent 2 is assigned a role that forbids him to select the action *Thriller* (say, he is under 12), then agent 1, assuming he knows the role of agent 2, can safely choose *Comedy* resulting in coordination. Note that there is only one equilibrium left in the subgame formed after removing the action *Thriller* from the action set of agent 2.

In general, suppose that there are n available roles (not necessarily distinct), that the state is fully observable to the agents, and that the following facts are common knowledge among agents:

For each agent i in parallel
    $I = \{\}$.
    For each role $j = 1, \ldots, n$
        Compute the potential r_{ij} of agent i for role j.
        Broadcast r_{ij} to all agents.
    End
    Wait until all $r_{i'j}$, for $j = 1, \ldots, n$, are received.
    For each role $j = 1, \ldots, n$
        Assign role j to agent $i^* = \arg\max_{i' \notin I}\{r_{i'j}\}$.
        Add i^* to I.
    End
End

FIGURE 4.2: Communication-based greedy role assignment

- There is a fixed ordering $\{1, 2, \ldots, n\}$ of the roles. Role 1 must be assigned first, followed by role 2, etc.

- For each role there is a function that assigns to each agent a 'potential' that reflects how appropriate that agent is for the specific role, given the current state. For example, the potential of a soccer robot for the role attacker can be given by its negative Euclidean distance to the ball.

- Each agent can be assigned only one role.

Then role assignment can be carried out, for instance, by a **greedy** algorithm in which each role (starting from role 1) is assigned to the agent that has the highest potential for that role, and so on until all agents have been assigned a role. When communication is not available, each agent can run this algorithm identically and in parallel, assuming that each agent can compute the potential of each other agent. When communication is available, an agent only needs to compute its own potentials for the set of roles, and then broadcast them to the rest of the agents. Next it can wait for all other potentials to arrive in order to compute the assignment of roles to agents as above. In the communication-based case, each agent needs to compute $O(n)$ (its own) potentials instead of $O(n^2)$ in the communication-free case, but this is compensated by the total number $O(n^2)$ of potentials that need to be broadcast and processed by the agents. Figure 4.2 shows the greedy role assignment algorithm when communication is available.

4.4   COORDINATION GRAPHS

As mentioned above, roles can facilitate the solution of a coordination game by reducing the action sets of the agents prior to computing the equilibria. However, computing equilibria in a subgame can still be a difficult task when the number of involved agents is large; recall that the joint action space is exponentially large in the number of agents. As roles reduce the size of the action sets, we also need a method that reduces the number of agents involved in a coordination game.

Guestrin et al. (2002a) introduced the **coordination graph** as a framework for solving large-scale coordination problems. A coordination graph allows for the decomposition of a coordination game into several smaller subgames that are easier to solve. Unlike roles where a single subgame is formed by the reduced action sets of the agents, in this framework various subgames are formed, each typically involving a small number of agents.

In order for such a decomposition to apply, the main assumption is that the global payoff function $u(a)$ can be written as a linear combination of k local payoff functions f_j, for $j = 1, \ldots, k$, each involving fewer agents. For example, suppose that there are $n = 4$ agents, and $k = 3$ local payoff functions, each involving two agents:

$$u(a) = f_1(a_1, a_2) + f_2(a_1, a_3) + f_3(a_3, a_4). \qquad (4.1)$$

Here, for instance $f_2(a_1, a_3)$ involves only agents 1 and 3, with their actions a_1 and a_3. Such a decomposition can be graphically represented by a graph (hence the name), where each node represents an agent and each edge corresponds to a local payoff function. For example, the decomposition (4.1) can be represented by the graph of Fig. 4.3.

Many practical problems can be modeled by such additively decomposable payoff functions. For example, in a computer network nearby servers may need to coordinate their actions in order to optimize overall network traffic; in a firm with offices in different cities, geographically nearby offices may need to coordinate their actions in order to maximize global sales; in a soccer team, nearby players may need to coordinate their actions in order to improve team performance; and so on.

Let us now see how this framework can be used for coordination. A solution to the coordination problem is by definition a Pareto optimal Nash equilibrium in the corresponding strategic game, that is, a joint action a^* that maximizes $u(a)$. We will describe two solution

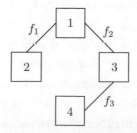

FIGURE 4.3: A coordination graph for a four-agent problem

methods: an exact one that is based on variable elimination, and an approximate one that is based on message passing.

4.4.1   Coordination by Variable Elimination

The linear decomposition of $u(a)$ in a coordination graph allows for the computation of a^* by a sequential maximization procedure, called **variable elimination**, in which agents are eliminated one after the other. We will illustrate this method on the above example. We start by eliminating agent 1 in (4.1). We collect all local payoff functions that involve agent 1; these are f_1 and f_2. The maximum of $u(a)$ can then be written

$$\max_a u(a) = \max_{a_2,a_3,a_4} \left\{ f_3(a_3, a_4) + \max_{a_1} \left[f_1(a_1, a_2) + f_2(a_1, a_3) \right] \right\}. \tag{4.2}$$

Next we perform the inner maximization over the actions of agent 1. For each combination of actions of agents 2 and 3, agent 1 must choose an action that maximizes $f_1 + f_2$. This essentially involves computing the best-response function $B_1(a_2, a_3)$ of agent 1 (see Section 3.4) in the subgame formed by agents 1, 2, and 3, and the sum of payoffs $f_1 + f_2$. The function $B_1(a_2, a_3)$ can be thought of as a conditional strategy for agent 1, given the actions of agents 2 and 3.

The above maximization and the computation of the best-response function of agent 1 define a new payoff function $f_4(a_2, a_3) = \max_{a_1}[f_1(a_1, a_2) + f_2(a_1, a_3)]$ that is independent of a_1. Agent 1 has now been eliminated. The maximum (4.2) becomes

$$\max_a u(a) = \max_{a_2,a_3,a_4} \left[f_3(a_3, a_4) + f_4(a_2, a_3) \right]. \tag{4.3}$$

We can now eliminate agent 2 as we did with agent 1. In (4.3), only f_4 involves a_2, and maximization of f_4 over a_2 gives the best-response function $B_2(a_3)$ of agent 2 which is a function of a_3 only. This in turn defines a new payoff function $f_5(a_3)$, and agent 2 is eliminated. Now we can write

$$\max_a u(a) = \max_{a_3,a_4} \left[f_3(a_3, a_4) + f_5(a_3) \right]. \tag{4.4}$$

Agent 3 is eliminated next, resulting in $B_3(a_4)$ and a new payoff function $f_6(a_4)$. Finally, $\max_a u(a) = \max_{a_4} f_6(a_4)$, and since all other agents have been eliminated, agent 4 can simply choose an action a_4^* that maximizes f_6.

The above procedure computes an optimal action only for the last eliminated agent (assuming that the graph is connected). For the other agents it computes only conditional strategies. A second pass in the reverse elimination order is needed so that all agents compute their optimal (unconditional) actions from their best-response functions. In the above example, plugging a_4^* into $B_3(a_4)$ gives the optimal action a_3^* of agent 3. Similarly, we get a_2^* from $B_2(a_3^*)$ and a_1^* from $B_1(a_2^*, a_3^*)$, which gives us the optimal joint action $a^* = (a_1^*, a_2^*, a_3^*, a_4^*)$. Note that

For each agent in parallel
    $F = \{f_1, \ldots, f_k\}$.
    For each agent $i = 1, 2, \ldots, n$
        Find all $f_j(a_i, a_{-i}) \in F$ that involve a_i.
        Compute $B_i(a_{-i}) = \arg \max_{a_i} \sum_j f_j(a_i, a_{-i})$.
        Compute $f_{k+i}(a_{-i}) = \max_{a_i} \sum_j f_j(a_i, a_{-i})$.
        Remove all $f_j(a_i, a_{-i})$ from F and add $f_{k+i}(a_{-i})$ in F.
    End
    For each agent $i = n, n-1, \ldots, 1$
        Choose $a_i^* \in B_i(a_{-i}^*)$ based on a fixed ordering of actions.
    End
End

FIGURE 4.4: Communication-free variable elimination

one agent may have more than one best-response actions, in which case the first action can be chosen according to an a priori ordering of the actions of each agent that must be common knowledge.

The complete algorithm, which we will refer to as coordination by variable elimination, is shown in Fig. 4.4. Note that the notation $-i$ that appears in $f_j(a_i, a_{-i})$ refers to all agents other than agent i that are involved in f_j, and it does not necessarily include all $n-1$ agents. Similarly, in the best-response functions $B_i(a_{-i})$ the action set a_{-i} may involve less than $n-1$ agents. The algorithm runs identically for each agent in parallel. For that we require that all local payoff functions are common knowledge among agents, and that there is an a priori ordering of the action sets of the agents that is also common knowledge. The latter assumption is needed so that each agent will finally compute the same joint action. The main advantage of this algorithm compared to coordination by social conventions is that here we need to compute best-response functions in subgames involving only few agents, as opposed to computing best-response functions in the complete game involving all n agents.

For simplicity, in the above algorithm we have fixed the elimination order of the agents as $1, 2, \ldots, n$. However, this is not necessary; each agent running the algorithm can choose a different elimination order, and the resulting joint action a^* will always be the same. The total runtime of the algorithm, however, will not be the same; different elimination orders produce different intermediate payoff functions, and thus subgames of different size. It turns out that computing the elimination order that minimizes the execution time of the algorithm is a hard (NP-complete) problem (Arnborg et al., 1987). A good heuristic is to eliminate agents that have the fewest neighbors.

When communication is available, we do not need to assume that all local payoff functions f_j are common knowledge and that the actions are ordered. In the forward pass, each agent can maintain in its local memory the payoff functions that involve only this agent. The initial distribution of payoff functions to the agents can be done as follows: agent 1 in the elimination order takes all payoff functions that involve this agent, agent 2 takes all functions that involve this agent and are not distributed to agent 1, and so on, until no more payoff functions are left. When an agent computes its best-response function and generates a new payoff function, the agent can broadcast this function to the other agents involved in it. In fact, the agent needs to send the payoff function only to the first non-eliminated agent whose action appears in the domain of this function. Similarly, in the backward pass an agent can wait for the optimal actions of the other agents (unless it is the last eliminated agent), then choose any action from its best-response action set, and finally broadcast this action to all agents that need it in their best-response functions. Figure 4.5 shows the communication-based version of variable elimination.

A crucial difference between the algorithms of Figs. 4.4 and 4.5 is that in the communication-based case the elimination order of the agents must be fixed a priori and it must be common knowledge among the agents. In terms of complexity, the forward pass is slightly slower than in the communication-free case, because here the generated payoffs need to be communicated to the other involved agents. On the other hand, when communication is available, the backward pass can be fully asynchronous. One can also think of asynchronous versions of the forward pass in which many agents are simultaneously eliminated. This would require some additional book-keeping for storing the pairwise dependencies between agents.

<u>For</u> each agent i in parallel
    <u>If</u> $i \neq 1$
        Wait until agent $i - 1$ sends OK.
    <u>End</u>
    Let $f_j(a_i, a_{-i})$ be all local payoff functions (initial and communicated) that involve agent i.
    Compute $B_i(a_{-i}) = \arg\max_{a_i} \sum_j f_j(a_i, a_{-i})$.
    Compute $f^*(a_{-i}) = \max_{a_i} \sum_j f_j(a_i, a_{-i})$.
    Send $f^*(a_{-i})$ to agent $j = \min\{i + 1, \ldots, n\}$, $j \in -i$.
    <u>If</u> $i \neq n$
        Send OK to agent $i + 1$.
        Wait until all a^*_{-i} are received.
    <u>End</u>
    Choose any $a^*_i \in B_i(a^*_{-i})$.
    Broadcast a^*_i to all agents j such that $a_i \in \text{domain}(B_j)$.
<u>End</u>

FIGURE 4.5: Communication-based variable elimination

4.4.2 Coordination by Message Passing

Variable elimination is an exact method (it always computes an optimal joint action), but it suffers from two limitations. First, for densely connected graphs the runtime of the method can be exponential in the number of agents (for example, a particular elimination order may cause the graph to become fully connected). Second, variable elimination can only produce a solution after the end of its backward pass, which can be unacceptable when real-time behavior is in order: often, decision making is done under time constraints, and there is a deadline after which the payoff of the team becomes zero (think of a research team that try to a submit a proposal before a deadline). In such cases we would like to have an **anytime** algorithm that improves the quality of the solution over time and (if possible) eventually computes the optimal solution.

Such anytime behavior can be achieved by distributed algorithms that are based on **message passing**. Here we will describe one such algorithm, called **max-plus**, that was originally developed for computing maximum a posteriori (MAP) solutions in Bayesian networks (Pearl, 1988). In this algorithm, neighboring agents in the graph repeatedly send messages to each other, where a message is a local payoff function for the receiving agent. Suppose that we have a coordination graph that defines a payoff function as a sum of two-agent local payoff functions:

$$u(a) = \sum_{(i,j)} f_{ij}(a_i, a_j) \qquad (4.5)$$

where the summation is over all (i, j) pairs of neighboring agents in the graph. In each time step, each agent i sends a message μ_{ij} to a (randomly picked) neighbor j, where μ_{ij} is a local payoff function for the receiving agent j defined as

$$\mu_{ij}(a_j) = \max_{a_i} \left\{ f_{ij}(a_i, a_j) + \sum_{k \in \Gamma(i) \setminus j} \mu_{ki}(a_i) \right\} \qquad (4.6)$$

where $\Gamma(i) \setminus j$ denotes all neighbors of agent i except agent j. Messages are exchanged until they converge to a fixed point, or until some external signal stops the process. The two operators involved in (4.6), a maximization and a summation, give the name max-plus to the algorithm.

When the graph is cycle-free (tree), max-plus always converges after a finite number of steps to a fixed point in which the messages do not change anymore (Pearl, 1988, Wainwright et al., 2004). If we define local functions g_i, one for each agent i, as

$$g_i(a_i) = \sum_{j \in \Gamma(i)} \mu_{ji}(a_i) \qquad (4.7)$$

then we can show that at convergence holds

$$g_i(a_i) = \max_{\{a' \mid a_i' = a_i\}} u(a'). \qquad (4.8)$$

If each agent locally computes

$$a_i^* = \arg\max_{a_i} g_i(a_i) \qquad\qquad (4.9)$$

and each optimal action a_i^* is unique (for all i), then at convergence the globally optimal action $a^* = \arg\max_a u(a)$ is also unique and has elements $a^* = (a_i^*)$ computed by only local optimizations (each agent maximizes $g_i(a_i)$ separately). If the local a_i^* are not unique, an optimal joint action can still be computed by dynamic programming (Wainwright et al., 2004, sec. 3.1). Hence, max-plus allows the decomposition of a difficult global optimization problem ($a^* = \arg\max_a u(a)$) into a set of local optimization problems (4.9) that are much easier to solve.

When the graph contains cycles, there are no guarantees that max-plus will converge, nor that the local maximizers a_i^* from (4.9) will comprise a global maximum at any time step. However, max-plus can still be used as an approximate coordination algorithm, that produces very good results in practice, much faster than variable elimination (Kok and Vlassis, 2006).

Max-plus is effective and simple to implement, but it comes with few performance guarantees in general graphs. Other algorithms exist, based on **branch and bound** or **hill climbing**, that can provably converge to the optimal solution (Modi et al., 2005), or to a k-optimal solution in which no subset of k or fewer agents can jointly improve the global payoff (Pearce and Tambe, 2007, Zhang et al., 2005).

4.5   NOTES AND FURTHER READING

Multiagent coordination has its roots in decentralized optimal control methods (Sandell et al., 1978). Early AI approaches to multiagent coordination are the 'contract net protocol' of Smith (1980) where tasks are dynamically distributed among agents using a bidding mechanism (see also Chapter 6), and the 'partial global planning' algorithm of Durfee and Lesser (1987) and Decker and Lesser (1995) in which agents exchange and refine local plans in order to reach a common goal. Jennings (1996) gives an overview of early coordination techniques in distributed AI. The framework of 'joint intentions' of Cohen and Levesque (1991) provides a formal characterization of multiagent coordination through a model of joint beliefs and intentions of the agents. Social conventions were introduced by Shoham and Tennenholtz (1992), as constraints on the set of allowed actions of a single agent at a given state (similar to the definition of roles in Section 4.3). Boutilier (1996) extended the definition to include also constraints on the joint action choices of a group of agents, and proposed the idea of coordination by lexicographic ordering. The greedy algorithm for role assignment was proposed by Castelpietra et al. (2000). Gmytrasiewicz and Durfee (2001) analyze coordination and communication in a setting where the agents model the knowledge of each other recursively (see also Chapter 5). Coordination graphs are due to Guestrin et al. (2002a) who suggested the use of variable elimination for coordination. The max-plus algorithm on coordination graphs was suggested by Vlassis et al. (2004). Coordination on a coordination graph is essentially identical to a distributed constraint optimization problem (DCOP) (Modi et al., 2005, Yokoo, 2000), a particular version of constraint processing (Dechter, 2003).

CHAPTER 5

Partial Observability

In the previous chapters we assumed that the world state is fully observable to the agents. Here we relax this assumption and examine the case where parts of the state are hidden to the agents. In such a partially observable world an agent must always reason about his knowledge, and the knowledge of the others, prior to making decisions. We formalize the notions of knowledge and common knowledge in such domains, and describe the model of a Bayesian game for multiagent decision making under partial observability.

5.1 THINKING INTERACTIVELY

In order to act rationally, an agent must always reflect on what he knows about the current world state. As we saw in Chapter 2, if the state is fully observable, an agent can do pretty well without extensive deliberation. If the state is partially observable, however, the agent must first consider carefully what he knows and what he does not know before choosing an action.

In a multiagent system, partial observability forces a rational agent to think interactively, that is, to take into account the knowledge of the other agents in his decision making. In addition, an agent must consider what the other agents know about him, and also what they know about his knowledge. In the previous chapters we have often used the term **common knowledge** to refer to something that every agent knows, that every agent knows that every other agent knows, and so on. In this chapter we will define knowledge and common knowledge more formally, and illustrate some of their implications through examples.

Partial observability may have various consequences to the decision making of the agents. For instance, optimal planning under partial observability can be a hard problem even in the single-agent case (Papadimitriou and Tsitsiklis, 1987). In the multiagent case, optimal planning under partial observability is provably intractable (Bernstein et al., 2002). The latter is due to the fact that, as stated above, each agent must take into account the knowledge of each other agent in its decision making, which can significantly increase the complexity of the problem. Later in this chapter we will see how the model of a **Bayesian game** can be used for multiagent decision making under partial observability.

5.2   INFORMATION AND KNOWLEDGE

In this section we will illustrate the concepts of information and common knowledge by means of a classical puzzle, the **puzzle of the hats**:

> Three agents (say, girls) are sitting around a table, each wearing a hat. A hat can be either red or white, but suppose that all agents are wearing red hats. Each agent can see the hat of the other two agents, but she does not know the color of her own hat. A person who observes all three agents asks them in turn whether they know the color of their hats. Each agent replies negatively. Then the person announces 'At least one of you is wearing a red hat', and then asks them again in turn. Agent 1 says *No*. Agent 2 also says *No*. But when he asks agent 3, she says *Yes*.

How is it possible that agent 3 can finally figure out the color of her hat? Before the announcement that at least one of them is wearing a red hat, no agent is able to tell her hat color. What changes then after the announcement? Seemingly the announcement does not reveal anything new; each agent already knows that there is at least one red hat because she can see the red hats of the other two agents.

Given that everyone has heard that there is at least one red hat, agent 3 can tell her hat color by reasoning as follows: 'Agent's 1 *No* implies that either me or agent 2 is wearing a red hat. Agent 2 knows this, so if my hat had been white, agent 2 would have said *Yes*. But agent 2 said *No*, so my hat must be red.'

Although each agent already knows (by perception) the fact that at least one agent is wearing a red hat, the key point is that the public announcement of the person *makes this fact common knowledge* among the agents. (Implicitly we have also assumed that it is common knowledge that each agent can see and hear well, and that she can reason rationally.) The puzzle is instructive as it demonstrates the implications of interactive reasoning and the strength of the common knowledge assumption.

Let us now try to formalize some of the concepts that appear in the puzzle. The starting point is that the world state is partially observable to the agents. Recall that in a partially observable world the perception of an agent provides only partial information about the true state by means of a deterministic or stochastic observation model (see Section 2.3). In the puzzle of the hats this model is a set-partition deterministic model, as we will see next.

Let S be the set of all states and $s \in S$ be the current (true) state of the world. We assume that the perception of an agent i provides information about the state s through an **information function** $P_i : S \mapsto 2^S$ that maps s to $P_i(s)$, a nonempty subset of S called the **information set** of agent i in state s. The interpretation of the information set is that when the true state is s, agent i thinks that any state in $P_i(s)$ can be the true state. The set $P_i(s)$ will always

		World states						
		a	b	c	d	e	f	g	h
	1	R	R	R	R	W	W	W	W
Agents	2	R	R	W	W	R	R	W	W
	3	R	W	R	W	R	W	R	W

FIGURE 5.1: The eight world states in the puzzle of the hats

contain s, but essentially this is the only thing that agent i knows about the true state. In the case of multiple agents, each agent can have a different information function.

In the puzzle of the hats, a state is a three-component vector containing the colors of the hats. Let R and W denote red and white. There are in total eight states $S = \{a, b, c, d, e, f, g, h\}$, as shown in Fig. 5.1. By assumption, the true state is $s = a$. From the setup of the puzzle we know that the state is partially observable to each agent; only two of the three hat colors are directly perceivable by each agent. In other words, in any state s the information set of each agent contains two equiprobable states, those in which the only difference is in her own hat color. For instance, in state $s = a$ the information set of agent 2 is $P_2(s) = \{a, c\}$, a two-state subset of S.

As we mentioned above, the information set $P_i(s)$ of an agent i contains those states in S that agent i considers possible if the true state is s. In general, we assume that the information function of an agent divides the state space into a collection of mutually disjoint subsets, called **cells**, that together form a **partition** $\mathcal{P}_i$ of S. The information set $P_i(s)$ for agent i in true state s is exactly that cell of $\mathcal{P}_i$ that contains s, while the union of all cells in $\mathcal{P}_i$ is S.

Based on the information functions, we can compute the partitions of the agents in the puzzle of the hats:

$$\mathcal{P}_1^t = \{\{a, e\}, \{b, f\}, \{c, g\}, \{d, h\}\} \tag{5.1}$$
$$\mathcal{P}_2^t = \{\{a, c\}, \{b, d\}, \{e, g\}, \{f, h\}\} \tag{5.2}$$
$$\mathcal{P}_3^t = \{\{a, b\}, \{c, d\}, \{e, f\}, \{g, h\}\} \tag{5.3}$$

where t refers to the time step before any announcement took place. Clearly, in the true state $s = a = RRR$ no agent knows her hat color, since the corresponding cell of each partition contains two equiprobable states. Thus, agent 1 considers a and e possible, agent 2 considers a and c possible, and agent 3 considers a and b possible. (Note again that we know that the true state is a but the agents in our puzzle do not.)

Now we make the additional assumption that all partitions are common knowledge among the agents. In the case of homogeneous agents, for instance, this is not an unrealistic assumption; typically each agent will be aware of the perception capabilities of each other. In the puzzle of the hats, for example, it is reasonable to assume that all agents can see and hear well and that they are all rational. Then, simply the positioning of the agents around the table makes the above partitions common knowledge.

If the partitions are common knowledge, then in state a agent 1 thinks that agent 2 may think that agent 3 might think that $h = WWW$ is possible! Why is that? Note from (5.1) that in state a agent 1 thinks that either a or e could be the true state. But if e is the true state, then from (5.2) we see that agent 2 may consider g to be the true state. But then we see from (5.3) that agent 3 may consider h to be the true state. Note how the above analytical framework allows for a fairly straightforward formulation of otherwise complicated statements.

Now the announcement of the person reveals that the true state is not h. This automatically changes the partitions of the agents:

$$\mathcal{P}_1^{t+1} = \{\{a, e\}, \{b, f\}, \{c, g\}, \{d\}, \{h\}\}$$
$$\mathcal{P}_2^{t+1} = \{\{a, c\}, \{b, d\}, \{e, g\}, \{f\}, \{h\}\}$$
$$\mathcal{P}_3^{t+1} = \{\{a, b\}, \{c, d\}, \{e, f\}, \{g\}, \{h\}\}.$$

(5.4)

Note that h has been disambiguated from d, f, and g, in the three partitions. The person then asks each agent in turn whether she knows the color of her hat. Agent 1 says *No*. In which case would agent 1 have said *Yes*? As we see from the above partitions, only in state d would agent 1 have known her hat color. But the true state is a, and in this state agent 1 still considers e possible.

The reply of agent 1 eliminates state d from the set of candidate states. This results in a refinement of the partitions of agents 2 and 3:

$$\mathcal{P}_1^{t+2} = \{\{a, e\}, \{b, f\}, \{c, g\}, \{d\}, \{h\}\}$$
$$\mathcal{P}_2^{t+2} = \{\{a, c\}, \{b\}, \{d\}, \{e, g\}, \{f\}, \{h\}\}$$
$$\mathcal{P}_3^{t+2} = \{\{a, b\}, \{c\}, \{d\}, \{e, f\}, \{g\}, \{h\}\}.$$

(5.5)

Next agent 2 is asked. From her partition P_2^{t+2} we see that she would have known her hat color only in state b or f (d and h are already ruled out by the previous announcements). However, in the true state a agent 2 still considers c possible, therefore she replies negatively. Her reply excludes b and f from the set of candidate states, resulting in a further refinement of the partitions of agent 1 and 3:

$$\mathcal{P}_1^{t+3} = \{\{a, e\}, \{b\}, \{f\}, \{c, g\}, \{d\}, \{h\}\}$$
$$\mathcal{P}_2^{t+3} = \{\{a, c\}, \{b\}, \{d\}, \{e, g\}, \{f\}, \{h\}\}$$
$$\mathcal{P}_3^{t+3} = \{\{a\}, \{b\}, \{c\}, \{d\}, \{e\}, \{f\}, \{g\}, \{h\}\}.$$

(5.6)

The partition of agent 3 now contains only singleton cells, thus agent 3 can now tell her hat color. Note that agents 1 and 2 still cannot tell their hat colors. In fact, they will be unable to tell their hat colors no matter how many more announcements will take place; the partitions (5.6) cannot be further refined. Interestingly, the above analysis would have been exactly the same if the true state had been any one in the set $\{a, c, e, g\}$. (Try to verify this with logical reasoning.)

5.3 COMMON KNOWLEDGE

Any subset E of S is called an **event**. If for an agent i holds $P_i(s) \subseteq E$ in true state s, then we say that agent i **knows**[1] E. Generalizing, the **knowledge function** of an agent i is defined as

$$K_i(E) = \{s \in S : P_i(s) \subseteq E\}. \tag{5.7}$$

That is, for any event E, the set $K_i(E)$ contains all states in which agent i knows E. It is not difficult to see that $K_i(E)$ can be written as the union of all cells of $\mathcal{P}_i$ that are fully contained in E. In the puzzle of the hats, for example, in the final partitions (5.6) holds $K_1(\{a, e, c\}) = \{a, e\}$, while for the event $E = \{a, c, e, g\}$ holds $K_i(E) = E$ for all $i = 1, 2, 3$.

An event $E \subseteq S$ is called **self-evident** to agent i if E can be written as a union of cells of $\mathcal{P}_i$. For example, in (5.6) the event $E = \{a, c, e, g\}$ is self-evident to all three agents. As another example, suppose that the state space consists of the integer numbers from 1 to 8, the true state is $s = 1$, and two agents have the following partitions:

$$\begin{aligned}
\mathcal{P}_1 &= \{\{1, 2\}, \{3, 4, 5\}, \{6\}, \{7, 8\}\} \\
\mathcal{P}_2 &= \{\{1, 2, 3\}, \{4\}, \{5\}, \{6, 7, 8\}\}.
\end{aligned} \tag{5.8}$$

In $s = 1$ agent 1 thinks that $\{1, 2\}$ are possible. Agent 1 also thinks that agent 2 may think that $\{1, 2, 3\}$ are possible. Furthermore, agent 1 thinks that agent 2 may think that agent 1 might think that $\{1, 2\}$ or $\{3, 4, 5\}$ are possible. But nobody needs to think beyond 5. In this example, the event $\{1, 2, 3, 4\}$ is self-evident to agent 2, while the event $\{1, 2, 3, 4, 5\}$ is self-evident to both agents.

We can now formalize the notion of common knowledge. For simplicity, the first definition is formulated for only two agents.

Definition 5.1. *An event $E \subseteq S$ is common knowledge between agents 1 and 2 in true state $s \in S$, if s is a member of every set in the infinite sequence $K_1(E)$, $K_2(E)$, $K_1(K_2(E))$, $K_2(K_1(E))$, ...*

Definition 5.2. *An event $E \subseteq S$ is common knowledge among a group of agents in true state $s \in S$, if s is a member of some set $F \subseteq E$ that is self-evident to all agents.*

---

[1]This definition of knowledge is related to the one used in **epistemic logic**. There an agent is said to know a fact ϕ if ϕ is true in all states the agent considers possible. In the event-based framework, an agent knows an event E if all the states the agent considers possible are contained in E. Fagin et al. (1995, sec. 2.5) show that the two approaches, logic-based and event-based, are equivalent.

It turns out that the two definitions are equivalent (Osborne and Rubinstein, 1994, prop. 74.2). However, the second definition is much easier to apply; it only requires computing self-evident sets that are unions of partition cells and thus easy to find. For instance, in the above example the event $E = \{1, 2, 3, 4, 5\}$ is common knowledge between the two agents because E is self-evident to both of them and the true state $s = 1$ belongs to E. Similarly, in the puzzle of the hats, in the final partitions (5.6), and with true state $s = a$, the event $E = \{a, c, e, g\}$ is common knowledge among all three agents.

5.4   PARTIAL OBSERVABILITY AND ACTIONS

So far we have discussed how the observations of the agents are related to the world states through the information functions, and what it means to say that an event is common knowledge among a group of agents. In this section we describe a framework that allows the agents to take rational decisions under partial observability.

Let us consider the puzzle of the hats again, and in particular the time step after the public announcement of the person. Below we identify the primitives of the multiagent interaction at that step that are relevant for the decision making of the agents.

5.4.1   States and Observations

The true state is $s = a = RRR$ and it is partially observable to the agents: each agent i receives an observation $\theta_i \in \Theta_i$ that provides information about s via the information function P_i. (In the previous sections of this chapter the observations θ_i were not explicit; here we make them explicit by associating each information set $P_i(s)$ with a corresponding observation θ_i.) For example, at $s = RRR$ agent 1 observes $\theta_1 = RR$, meaning that she sees the two red hats of agents 2 and 3, where θ_1 is a member of the set $\Theta_1 = \{RR, RW, WR, WW\}$ (all possible observations of agent 1). The profile of the individual observations of all agents (θ_i) defines the joint observation θ.

5.4.2   Observation Model

The partition model associates with each observation θ_i of agent i a single information set $P_i(s)$ that is a subset of the state space. For instance, $\theta_1 = RR$ is associated with the information set $P_1(s) = \{a, e\}$. In this problem each observation is a deterministic function of the state: the observation of each agent at each state is fully determined by the setup of the problem (the position of a girl around the table). As we mentioned in Section 2.3, more general observation models can be defined in which the coupling between states and observations is stochastic. For instance, an observation model could define a joint probability distribution $p(s, \theta)$ over states and joint observations, from which various other quantities can be computed, like $p(\theta)$ or $p(\theta|s)$, by using the laws of probability theory.[2]

5.4.3   Actions and Policies

In the puzzle of the hats each agent replies *Yes* or *No* to the question of the person 'Do you know your hat color?'. Such a reply can be regarded as an action taken by the agent given her current information. For example, in the final partitions of (5.6), agent 1 will reply *No* given her information set $\{a, e\}$ and agent 3 will reply *Yes* given her information set $\{a\}$. In general, in multiagent decision making under partial observability, the policy of each agent i is a mapping $\pi_i : \Theta_i \mapsto A_i$ from individual observations θ_i to individual actions $a_i = \pi_i(\theta_i)$. (Recall that this was the definition of a memoryless policy of a reflex agent in Section 2.2.) The profile of individual policies (π_i) defines the joint policy π.

5.4.4   Payoffs

In the puzzle of the hats the agents reply truthfully to the questions of the person. Although we have not explicitly defined a payoff function in this problem, we can think of an implicit payoff function that the agents maximize, in which, say, truthfulness is highly valued. In general, multiagent decision making requires defining an explicit payoff function Q_i for each agent. This function can take several forms; for instance, it can be a function $Q_i(s, a)$ over states and joint actions; or a function $Q_i(\theta, a)$ over joint observations and joint actions; or a function $Q_i(\theta_i, a)$ over individual observations and joint actions (we will see an example of such a function in Chapter 6). Note that often one form can be derived from the other; for instance, when an inverse observation model $p(s|\theta)$ is available, we can write $Q_i(\theta, a) = \sum_{s \in S} p(s|\theta) Q_i(s, a)$.

When the above primitives are defined, multiagent decision making under partial observability can be modeled by a **Bayesian game**, also known as *strategic game with imperfect information*. This is a combination of the strategic game model of Section 3.2 with the concepts of knowledge and partial observability defined in this chapter. In particular, a Bayesian game assumes that there is a set of states S, from which one state (the true state) is realized at the start of the game. The true state is only partially observable by the agents; each agent i receives an observation θ_i, also called the **type** of agent i, that is hidden to the other agents, and that is related to the state via a deterministic or stochastic observation model. Each agent additionally possesses a payoff function Q_i as described above. The solution of the game is a profile of individual policies $\pi_i(\theta_i)$ that are optimal according to some solution concept, for instance, Nash equilibrium (defined below). Note that each individual policy $\pi_i(\theta_i)$ specifies an action to take by agent i *for each* of his observations, and not only for the observation that the agent

---

[2] $p(A) = \sum_B p(A, B)$, and $p(A|B) = p(A, B)/p(B)$.

actually receives after the game has started. Such an *ex ante* solution to the game is necessary, as it encompasses the interactive-thinking idea that an agent i may be uncertain about what another agent j believes that i will play after observing some θ_i.

Depending on the type of observation model and payoff functions, different models of a Bayesian game exist. We will describe two such models. The first one assumes payoff functions defined over states and joint actions, in the form $Q_i(s, a)$, and additionally that each agent i has access to an inverse observation model that is conditional on individual observations, in the form $p(s|\theta_i)$. In this model, a Nash equilibrium is defined as follows:

Definition 5.3. *A **Nash equilibrium of a Bayesian game** is a Nash equilibrium of a new strategic game in which each player is a pair (agent i, observation θ_i) and has payoff function*

$$u_i(\pi_i(\theta_i)) = \sum_s p(s|\theta_i) Q_i(s, [\pi_i(\theta_i), a_{-i}(s)]) \tag{5.9}$$

where $a_{-i}(s)$ is the profile of actions taken by all other players except player (i, θ_i) at state s.

Clearly, in order for this definition to be applicable, each agent must be able to infer the action of each other agent at each state. This requires that the observation model is common knowledge, and that it is a deterministic model where, for each i, the observation θ_i is a deterministic function of s (for instance a partitional model as in the puzzle of the hats). In this case, the policy $\pi_j(\theta_j)$ of an agent j uniquely identifies his action at s through $a_j(s) = \pi_j(\theta_j(s))$.

The second model of a Bayesian game is not making use of states. Instead it assumes that payoffs are defined over joint observations and actions, in the form $Q_i(\theta, a)$, and that a marginal observation model $p(\theta)$ is available. In this case, a Nash equilibrium is defined as in Definition 5.3 with (5.9) replaced by

$$u_i(\pi_i(\theta_i)) = \sum_{\theta_{-i}} p(\theta_{-i}|\theta_i) Q_i(\theta, [\pi_i(\theta_i), \pi_{-i}(\theta_{-i})]) \tag{5.10}$$

where now the quantities $\pi_{-i}(\theta_{-i})$ are directly available, and $p(\theta_{-i}|\theta_i)$ can be computed from $p(\theta)$. This second model of a Bayesian game is easier to work with, and it is often preferred over the first one in practical problems.

In the special case of n collaborative agents with common payoff functions $Q_1 = \ldots = Q_n \equiv Q$, coordination requires computing a Pareto optimal Nash equilibrium (see Chapter 4). In the second model of a Bayesian game described above, such an equilibrium can be computed by the following:

		θ_2		$\bar{\theta}_2$	
		a_2	$\bar{a}_2$	a_2	$\bar{a}_2$
θ_1	a_1	+0.1	+2.2	+0.4	− 0.2
	$\bar{a}_1$	− 0.5	+2.0	+1.0	+2.0
$\bar{\theta}_1$	a_1	+0.4	− 0.2	+0.7	− 2.6
	$\bar{a}_1$	+1.0	+2.0	+2.5	+2.0

FIGURE 5.2: A Bayesian game with common payoffs involving two agents and binary actions and observations. The shaded entries indicate the Pareto optimal Nash equilibrium of this game.

Proposition 5.1. A Pareto optimal Nash equilibrium for a Bayesian game with a common payoff function $Q(\theta, a)$ is a joint policy $\pi^* = (\pi_i^*)$ that satisfies

$$\pi^* = \arg\max_{\pi} \sum_{\theta} p(\theta) Q(\theta, \pi(\theta)). \qquad (5.11)$$

Proof. From the perspective of some agent i, the above formula reads

$$\pi_i^* = \arg\max_{\pi_i} \sum_{\theta_i} p(\theta_i) \sum_{\theta_{-i}} p(\theta_{-i}|\theta_i) Q_i(\theta, [\pi_i(\theta_i), \pi_{-i}^*(\theta_{-i})]). \qquad (5.12)$$

A sum of terms is maximized when each of the terms is maximized, so there must hold

$$\pi_i^*(\theta_i) = \arg\max_{\pi_i(\theta_i)} \sum_{\theta_{-i}} p(\theta_{-i}|\theta_i) Q_i(\theta, [\pi_i(\theta_i), \pi_{-i}^*(\theta_{-i})]) \qquad (5.13)$$

which is the definition of a Nash equilibrium from (5.10). This shows that π^* is a Nash equilibrium. The proof that π^* is also Pareto optimal is left as an exercise. $\square$

Figure 5.2 shows an example of a two-agent Bayesian game with common payoffs, where each agent i has two available actions, $A_i = \{a_i, \bar{a}_i\}$, and two available observations, $\Theta_i = \{\theta_i, \bar{\theta}_i\}$. Assuming uniform $p(\theta)$, we can compute from (5.11) the Pareto optimal Nash equilibrium $\pi^* = (\pi_1^*, \pi_2^*)$ of the game, which is

$$\pi_1^* : \quad \pi_1^*(\theta_1) = \bar{a}_1, \quad \pi_1^*(\bar{\theta}_1) = \bar{a}_1 \qquad (5.14)$$
$$\pi_2^* : \quad \pi_2^*(\theta_2) = \bar{a}_2, \quad \pi_2^*(\bar{\theta}_2) = \bar{a}_2. \qquad (5.15)$$

This solution gives to each agent expected payoff $u_i = 2$.

5.5 NOTES AND FURTHER READING

The concept of common knowledge was introduced by Lewis (1969). Osborne and Rubinstein (1994, ch. 5) and Geanakoplos (1992) give good accounts on the topic, with several examples. The Definition 5.2 of common knowledge is due to Aumann (1976). Fagin et al. (1995) provide an epistemic-logic treatment of knowledge and common knowledge, and give several impossibility results in the case of unreliable communication between agents. One can also define common knowledge of actions, giving rise to a family of 'agreement' theorems (the validity of which has often been criticized); Samet (2006) provides a thoughtful approach to the problem. The model of a Bayesian game was introduced by Harsanyi (1967). Osborne (2003, ch. 9) provides a detailed exposition of Bayesian games with many examples. Among several applications, Bayesian games have been used as models for multiagent planning under partial observability (Hansen et al., 2004, Emery-Montemerlo et al., 2005, Oliehoek and Vlassis, 2007).

CHAPTER 6

Mechanism Design

In this chapter we study the problem of mechanism design, which is the development of agent interaction protocols that explicitly take into account the fact that the agents may be self-interested. We discuss the revelation principle and the Vickrey–Clarke–Groves (VCG) mechanism that allows us to build successful protocols in a variety of cases.

6.1   SELF-INTERESTED AGENTS

In the previous chapters we saw several examples of multiagent systems that consist of collaborative agents. The fact that the agents in such systems must collaborate for a common goal allows the development of algorithms, like the coordination algorithms of Chapter 4, in which the agents are assumed to be truthful to each other and behave as instructed. A soccer robot, for instance, would never violate a role assignment protocol like the one in Fig. 4.2, as this could potentially harm the performance of its team.

In many practical applications, however, we have to deal with **self-interested** agents, for instance agents that act on behalf of some owner who wants to maximize his or her own profit. A typical case is a software agent that participates in an electronic auction on the Internet. Developing an algorithm or protocol for such a system is a much more challenging task than in the collaborative case. First, we have to motivate an agent to participate in the protocol, which is not *a priori* the case. Second, we have to take into account the fact that an agent may try to *manipulate* the protocol for his own interest, leading to suboptimal results. The latter includes the possibility that the agent may lie, if needed.

The development of protocols that are stable (non-manipulable) and individually rational for the agents (no agent is worse off by participating) is the subject of **mechanism design** or **implementation theory**. As we will see next, a standard way to deal with the above two problems is to provide payments to the agents in exchange for their services.

6.2   THE MECHANISM DESIGN PROBLEM

In Chapter 3 we used the model of a strategic game to describe a situation in which a group of agents interact with each other. The primitives of such a game are the action sets A_i and the payoff functions $u_i(a)$ of the agents, for $i = 1, \ldots, n$, where $u_i(a)$ reflects the preference of agent i for the joint action a. Moreover, for any profile of payoff functions, a solution concept (for instance Nash equilibrium) allows us to make predictions over the set of outcomes that may result when the game is played. Similarly, in Chapter 5 we used the model of a Bayesian game to describe a situation where some primitives of the game are hidden to the agents. Our standpoint in Chapters 3 and 5 was that of an external observer who wants to know the outcome of a game, but cannot affect this outcome in any way.

In mechanism design we take a different stance. Here we assume a set $\mathcal{O}$ of possible **outcomes** over which a number of agents form preferences. Our task is to *design* a game that, when played by the agents, brings about a desired outcome from $\mathcal{O}$, for instance an outcome that is socially favorable by the agents. An outcome can be practically anything, for instance the assignment of an auction item or a network resource to an agent (see examples below). In this framework we[1] therefore use a game as a tool for achieving our design goals. The main difficulty, however, is that we often do not know the preferences of the agents in advance.

To model the individual preferences of the agents we use a Bayesian game formulation (see Section 5.4). We assume that each agent $i = 1, \ldots, n$ has some private information $\theta_i \in \Theta_i$, which defines the type of the agent, and which is not revealed to the other agents or to us. Moreover, we assume that the type of an agent fully specifies the preferences of this agent over the set of outcomes $o \in \mathcal{O}$. In particular, each agent i has a **valuation function** $v_i(\theta_i, o)$ that is parametrized on θ_i, such that agent i in type θ_i prefers outcome o to o' if and only if $v_i(\theta_i, o) > v_i(\theta_i, o')$. We assume that the valuation functions of all agents are common knowledge, but each individual type θ_i is only privately known to agent i.

In mechanism design we additionally assume the existence of a **social choice function** $f : \Theta \mapsto O$ that maps any profile $\theta = (\theta_i)$ of agent types to a desired outcome $o = f(\theta)$. We can think of f as an algorithm that solves an optimization problem: given n inputs θ_i, the function f computes an outcome o that maximizes a functional over the set of agents, valuations. For instance, an **allocatively efficient** social choice function will choose the outcome that maximizes the sum of the agents' valuations:

$$f(\theta) = \arg\max_{o \in \mathcal{O}} \sum_{i=1}^{n} v_i(\theta_i, o). \tag{6.1}$$

The function f is assumed to be common knowledge among the agents.

If we had access to all agents' types θ_i, then we could compute the desired optimal outcome simply by inserting $\theta = (\theta_i)$ in f in (6.1) (assuming of course that we have a tractable algorithm for doing this). However, as we saw above, θ_i is revealed only to agent i. One option

---

[1]In this chapter we will use 'he' to refer to an agent, and 'we' to refer to the mechanism designer.

is to ask each agent to tell us his type, but there is no guarantee that an agent will report his true type! Recall that each agent i forms his own preferences over outcomes, given by his valuation function $v_i(\theta_i, o)$ that is parametrized by his true type θ_i. If by reporting a false type $\tilde{\theta}_i \neq \theta_i$ an agent i expects to receive higher payoff than by reporting his true type θ_i, then this agent may certainly consider lying. For instance, if a social choice function chooses the outcome that is last in the preferences of agent 1, that is, $f(\theta) = \arg\min_o v_1(\theta_1, o)$, then agent 1 will report a false type $\tilde{\theta}_i$ for which $\arg\min_o v_1(\tilde{\theta}_1, o) = \arg\max_o v_1(\theta_1, o)$.

The challenge therefore is to design **non-manipulable** mechanisms in which no agent can benefit from not abiding by the rules of the mechanism. For instance, if a mechanism requires from each agent to report his true type, then we would like truth-telling to be indeed in the best interests of each agent. Viewed from a computational perspective, we can characterize mechanism design as the development of efficient and robust algorithms for optimization problems with distributed parameters, where these parameters are controlled by agents that have different preferences for different solutions.

We focus here on simple mechanisms in the form of a Bayesian game with the following primitives:

- A_i is the set of available actions of agent i.
- Θ_i is the set of types of agent i.
- $g : A \mapsto O$ is an **outcome function** that maps a joint action $a = (a_i)$ to an outcome $o = g(a)$.
- $Q_i(\theta_i, a)$ is the payoff function of agent i that is defined as

$$Q_i(\theta_i, a) = v_i(\theta_i, g(a)) + \xi_i(g(a)) \qquad (6.2)$$

where $\xi_i : O \mapsto \mathbb{R}$ are **payment functions**, so that agent i receives payment $\xi_i(o)$ when outcome o is selected.

Including payment functions ξ_i is essential because we need to motivate the agents to participate in the mechanism; as we mentioned above, participation for an agent is not *a priori* the case. A mechanism in which no agent is worse off by participating, that is, $Q_i(\theta_i, a) \geq 0$ for all i, θ_i, and a, is called **individually rational**.

When we fix the action sets A_i, the outcome function g, and the payment functions ξ_i, the above mechanism becomes a standard Bayesian game which we will denote by $\mathcal{M} = (A_i, g, \xi_i)$. When the agents are confronted with this game, they are expected to choose a profile of conditional policies $\pi^* = (\pi_i^*)$ according to some solution concept, as we explained in Section 5.4. When the true private types θ_i are realized to the agents, each agent i will take action $\pi_i^*(\theta_i)$ according to his true type θ_i. The resulting joint action $\pi^*(\theta) = (\pi_i^*(\theta_i))$ is then mapped through g to an outcome $g(\pi^*(\theta))$. Our task as mechanism designer is to ensure that the selected joint action brings about the desired outcome, that is, $g(\pi^*(\theta)) = f(\theta)$.

As solution concept we may consider a Nash equilibrium, as in Section 5.4. Alternatively we may consider the following:

Definition 6.1. *A joint policy $\pi^* = (\pi_i^*)$ in a Bayesian game is an* **equilibrium in dominant strategies** *if for every agent i and every type θ_i of agent i holds*

$$Q_i(\theta_i, [\pi_i^*(\theta_i), a_{-i}]) \geq Q_i(\theta_i, [a_i, a_{-i}]) \qquad (6.3)$$

for all joint actions (a_i, a_{-i}).

That is, each agent i chooses for each of his types θ_i the action $\pi_i^*(\theta_i)$ with the highest payoff. In particular, note that an agent i does not consider in the above equilibrium the types θ_{-i} and the policies $\pi_{-i}^*(\theta_{-i})$ of the other agents. This is in contrast to the solution concept of a Nash equilibrium (5.10) where each agent i is assumed to possess a conditional distribution $p(\theta_{-i}|\theta_i)$ over the types of the other agents, and must know the policies of the other agents at the equilibrium.

Our choice of such a solution concept is motivated by the fact that we would like to design mechanisms in which each agent can compute his optimal action without having to worry about the actions of the other agents. In terms of predictive power for the solutions of a game, an equilibrium in dominant actions is weaker than both a Nash equilibrium and an equilibrium computed by iterated elimination of strictly dominated actions (see Chapter 3). However, in the context of mechanism design, the existence of such an equilibrium guarantees that every (rational) agent will adhere to it, even if he has no information about the preferences of the other agents. Such an equilibrium solution is also very attractive computationally, because an agent does not need to consider the types or the policies of the other agents.

Summarizing, the mechanism design problem can be defined as follows:

Definition 6.2 (The mechanism design problem). *Given a set of outcomes $o \in \mathcal{O}$, a profile of valuation functions $v_i(\theta_i, o)$ parametrized by θ_i, and a social choice function $f(\theta)$, find appropriate action sets A_i, an outcome function $g(a)$, and payment functions $\xi_i(o)$, such that for any profile of true types $\theta = (\theta_i)$ and for payoff functions $Q_i(\theta_i, a)$ defined via (6.2) holds $g(\pi^*(\theta)) = f(\theta)$, where π^* is an equilibrium in dominant strategies of the Bayesian game $\mathcal{M} = (A_i, g, \xi_i)$. In this case we say that the mechanism $\mathcal{M}$* **implements** *the social choice function f in dominant strategies.*

6.2.1   Example: An Auction

Consider the following mechanism design problem (an auction). We have n agents and an item (for example, a resource in a computer network). We want to assign the item to the agent that values it most, but we do not know the true valuations (types) of the agents. In this example, an outcome $o \in \{1, \ldots, n\}$ is the index of the agent to whom the item is assigned, while the valuation function of an agent i with type $\theta_i \in \mathbb{R}^+$ is $v_i(\theta_i, o) = \theta_i$ if $o = i$ and zero otherwise. The social choice function is $f(\theta_1, \ldots, \theta_n) = \arg \max_i \{\theta_i\}$ which is a special case of (6.1). If we do not include a payment function, that is $\xi_i = 0$ for all i, then a mechanism $\mathcal{M}_1 = (A_i, g, \xi_i)$ that implements f is always individually rational because for an agent i holds $Q_i(\theta_i, \cdot) = v_i(\theta_i, \cdot)$ which is either $\theta_i > 0$ or zero.

6.3 THE REVELATION PRINCIPLE

Looking at Definition 6.2, mechanism design seems a formidable task. Our design options can in principle involve all possible action sets A_i, all possible outcome functions g, and all possible payment functions ξ_i that we could provide to the agents. Searching in the space of all $\mathcal{M} = (A_i, g, \xi_i)$ for a mechanism that implements f would be infeasible. Fortunately, there is a theorem that tells us that we do not need to search in the space of all possible mechanisms.

Proposition 6.1 (Revelation principle). If a social choice function f is implementable in dominant strategies by a mechanism $\mathcal{M} = (A_i, g, \xi_i)$, then f is also implementable in dominant strategies by a mechanism $\mathcal{M}' = (\Theta_i, f, \xi_i)$ in which each agent is simply asked to report his type. Moreover, the dominant strategy of each agent i in $\mathcal{M}'$ is to report his true type θ_i.

Proof. If f is implementable by $\mathcal{M}$ in dominant strategies, then $f(\theta) = g(\pi^*(\theta))$ for all θ, where $\pi^* = (\pi_i^*)$ is an equilibrium in dominant strategies in $\mathcal{M}$. The latter implies that for each agent i, each type θ_i of agent i, and each joint type $(\tilde{\theta}_i, \tilde{\theta}_{-i})$ holds

$$Q_i^{\mathcal{M}}(\theta_i, [\pi_i^*(\theta_i), \pi_{-i}^*(\tilde{\theta}_{-i})]) \geq Q_i^{\mathcal{M}}(\theta_i, [\pi_i^*(\tilde{\theta}_i), \pi_{-i}^*(\tilde{\theta}_{-i})]). \tag{6.4}$$

Since the ξ_i are identical in $\mathcal{M}$ and $\mathcal{M}'$, using (6.2) we can rewrite (6.4) as

$$Q_i^{\mathcal{M}'}(\theta_i, [\theta_i, \tilde{\theta}_{-i}]) \geq Q_i^{\mathcal{M}'}(\theta_i, [\tilde{\theta}_i, \tilde{\theta}_{-i}]) \tag{6.5}$$

which proves that truth-telling is a dominant strategy in $\mathcal{M}'$, and hence f is implementable in dominant strategies by $\mathcal{M}'$. $\qquad\square$

A mechanism in the form $\mathcal{M} = (\Theta_i, f, \xi_i)$ in which each agent is asked to report his type is called a **direct-revelation** mechanism. A direct-revelation mechanism in which truth-telling is a dominant strategy for every agent is called **strategy-proof**. The revelation principle is remarkable because it allows us to restrict our attention to strategy-proof mechanisms only. One of its consequences, for example, is that if we cannot implement a social choice function by a strategy-proof mechanism, then there is no way to implement this function in dominant strategies by *any* other general mechanism. The revelation principle has been a powerful theoretical tool for establishing several possibility and impossibility results in mechanism design; Parkes (2001) provides more details and references.

6.3.1 Example: Second-price Sealed-bid (Vickrey) Auction

Let us return to the auction example, and consider a direct-revelation mechanism $\mathcal{M}_2(\Theta_i, f, \xi_i)$ in which each agent i is asked to bid a price $\tilde{\theta}_i$, and the item is allocated to the agent with the highest bid. The winning agent must then pay tax (negative payment) equal to the second highest bid, that is $\xi_i = -\max_{j \neq i}\{\tilde{\theta}_j\}$, whereas the other agents do not have to pay anything. In this case, the payoff function of an agent i with true valuation θ_i equals $Q_i(\theta_i, \tilde{\theta}) = \theta_i + \xi_i$ if $f(\tilde{\theta}) = i$ and zero otherwise. Mechanism $\mathcal{M}_2$ is individually rational because for the winning agent k holds $Q_k(\theta_k, \cdot) = \theta_k + \xi_k \geq 0$, while for the other agents $j \neq k$ holds $Q_j(\theta_j, \cdot) = 0$.

We will now show that in mechanism $\mathcal{M}_2$ truth-telling is a dominant strategy for each agent, that is, each agent must bid his true valuation. The payoff of agent i when reporting $\tilde{\theta}_i$ is $Q_i(\theta_i, [\tilde{\theta}_i, \tilde{\theta}_{-i}]) = \theta_i + \xi_i$ if $\tilde{\theta}_i > -\xi_i$ and zero otherwise. Ignoring ties, if $\theta_i > -\xi_i$ then any bid $\tilde{\theta}_i > -\xi_i$ is optimal (results in positive payoff $Q_i(\theta_i, [\tilde{\theta}_i, \tilde{\theta}_{-i}]) = \theta_i + \xi_i > 0$). If $\theta_i < -\xi_i$ then any bid $\tilde{\theta}_i < -\xi_i$ is optimal (results in zero payoff). Truth-telling bid $\tilde{\theta}_i = \theta_i$ is optimal in both cases, and thus it is a dominant strategy in $\mathcal{M}_2$.

6.4 THE VICKREY–CLARKE–GROVES MECHANISM

The mechanism $\mathcal{M}_2$ in the auction example above is a strategy–proof mechanism that implements the social choice function $f(\theta_1, \ldots, \theta_n) = \arg\max_i\{\theta_i\}$ which is a special case of an allocatively efficient social choice function (6.1). We return now to the more general case. We assume a direct-revelation mechanism in which the agents are asked to report their types, and based on their reports $\tilde{\theta} = (\tilde{\theta}_i)$ the mechanism computes an optimal outcome $f(\tilde{\theta})$ that solves

$$f(\tilde{\theta}) = \arg\max_{o \in \mathcal{O}} \sum_{i=1}^{n} v_i(\tilde{\theta}_i, o). \tag{6.6}$$

In a **Groves** mechanism, the payment function that is associated with a profile of reported types $\tilde{\theta}$ is defined for each agent as

$$\xi_i(f(\tilde{\theta})) = \sum_{j \neq i} v_j(\tilde{\theta}_j, f(\tilde{\theta})) - h_i(\tilde{\theta}_{-i}) \tag{6.7}$$

for arbitrary function $h_i(\tilde{\theta}_{-i})$ that does not depend on the report of agent i. In this case, and for payoffs given by (6.2), we can show the following (the proof is left as an exercise):

Proposition 6.2. A Groves mechanism is a strategy–proof mechanism.

Having the freedom to choose any function $h_i(\tilde{\theta}_{-i})$, the **Clarke** mechanism, also known as **Vickrey–Clarke–Groves** (VCG) mechanism, uses

$$h_i(\tilde{\theta}_{-i}) = \sum_{j \neq i} v_j(\tilde{\theta}_j, f'(\tilde{\theta}_{-i})) \qquad (6.8)$$

where $f'(\tilde{\theta}_{-i})$ is an allocatively efficient social choice function with agent i excluded:

$$f'(\tilde{\theta}_{-i}) = \arg\max_{o \in \mathcal{O}} \sum_{j \neq i} v_j(\tilde{\theta}_j, o). \qquad (6.9)$$

Under quite general conditions, the VCG mechanism can be shown to be individually rational. Moreover, in some applications the payments ξ_i to the agents are negative, so the mechanism does not need to be externally subsidized (however, the collected tax must be burnt).

6.4.1 Example: Shortest Path

This is classical example with many applications, that is based on the VCG mechanism. We want to compute the shortest path between two fixed nodes in a graph. Each edge i in the graph has cost (length) $\theta_i \geq 0$, and is operated by an agent (say, a transportation company) who would preferably stay out of the path. We do not know the cost of each edge in advance, and we want to design a mechanism in which each agent reports his true cost.

Translated in the language of mechanism design, an outcome o is an ordered list of agents, indices (the edges that are included in the shortest path); agent i has type θ_i (the cost of his edge), and valuation function $v_i(\theta_i, o) = -\theta_i$ if $i \in o$ and zero otherwise; and the social choice function f is an algorithm (for instance Dijkstra's shortest path algorithm) that solves (6.6), that is, computes the shortest path given the reported costs.

A VCG mechanism solves the above problem by providing nonzero payments to all agents i that are included in a shortest path solution. These payments are computed from (6.7) and (6.8) as:

$$\xi_i(f(\tilde{\theta})) = \sum_{j \neq i} v_j(\tilde{\theta}_j, f(\tilde{\theta})) - \sum_{j \neq i} v_j(\tilde{\theta}_j, f'(\tilde{\theta}_{-i})) = \tilde{\theta}_i - C + C' \qquad (6.10)$$

where C is the additive cost (length) of the shortest path solution, and C' is the length of the shortest path solution after edge i is removed from the graph. From (6.2) and (6.10), the payoff of agent i under truth-telling is $Q_i(\theta_i, [\theta_i, \tilde{\theta}_{-i}]) = -\theta_i + \theta_i - C + C'$, which is always nonnegative since removing an edge from a graph can never generate a shorter path. It is therefore individually rational for an agent to participate in this mechanism, and because VCG mechanisms are strategy-proof, each agent will report his true cost.

6.5   NOTES AND FURTHER READING

More detailed expositions on the topic of mechanism design are provided by Osborne and Rubinstein (1994, ch. 10), Sandholm (1999), Parkes (2001), and Conitzer (2006). The papers of Vickrey (1961), Clarke (1971), and Groves (1973) are seminal. The revelation principle is due to Gibbard (1973). Computational issues in mechanism design are discussed among others by Nisan (1999), Parkes (2001), Conitzer (2006), and Papadimitriou (2001). The latter writes: 'All design problems [in computer science] are now mechanism design problems'. Parkes and Shneidman (2004) discuss distributed implementations of the VCG mechanism in which the optimization load is distributed over the agents. Conitzer (2006) and Conitzer and Sandholm (2007) describe alternative approaches to mechanism design in which a mechanism is automatically built for a given problem instance.

CHAPTER 7

Learning

In this chapter we briefly address the issue of learning, in particular reinforcement learning which allows agents to learn from delayed rewards. We outline existing techniques for single-agent systems, and show how they can be extended in the multiagent case.

7.1 REINFORCEMENT LEARNING

Reinforcement learning is a generic name given to a family of techniques in which an agent tries to learn a task by directly interacting with the environment. The method has its roots in the study of animal behavior under the influence of external stimuli (Thorndike, 1898). In the last two decades, reinforcement learning has been extensively studied in artificial intelligence, where the emphasis is on how agents can improve their performance in a given task by perception and trial-and-error. The field of single-agent reinforcement learning is mature, with well-understood theoretical results and many practical techniques (Bertsekas and Tsitsiklis, 1996, Sutton and Barto, 1998).

On the other hand, **multiagent reinforcement learning**, where several agents are simultaneously learning by interacting with the environment and with each other, is still an active area of research, with a mix of positive and negative results. The main difficulty in extending reinforcement learning to multiagent systems is that the dynamics of concurrently learning systems can be very complicated, which calls for different approaches to modeling and analysis than those used in single-agent systems.

In this chapter we will outline the theory and some standard algorithms for single-agent reinforcement learning, and then briefly discuss their multiagent extensions. We must unavoidably be laconic as the literature on the topic has grown large; the reader is referred to the book of Greenwald (2007) for a more detailed treatment.

7.2 MARKOV DECISION PROCESSES

In Chapter 2 we described a generic utility-based framework that allows an agent to behave optimally under conditions of uncertainty. In this section we describe a framework that allows an agent to *learn* optimal policies in a variety of tasks.

The sequential decision making of a single agent in an observable stochastic world with Markovian transition model is called a **Markov decision process** (MDP). A (finite) MDP is formally defined by the following elements:

- Discrete time $t = 0, 1, 2, \ldots$.

- A discrete set of states $s \in S$.

- A discrete set of actions $a \in A$.

- A stochastic transition model $p(s'|s, a)$, so that the world transitions stochastically to state s' when the agent takes action a at state s.

- A **reward** function $R: S \times A \mapsto \mathbb{R}$, so that the agent receives reward $R(s, a)$ when it takes action a at state s.

- A planning horizon, which can be infinite.

The task of the agent is to maximize a function of accumulated reward over its planning horizon. A standard such function is the **discounted future reward** $R(s_t, a_t) + \gamma R(s_{t+1}, a_{t+1}) + \gamma^2 R(s_{t+2}, a_{t+1}) + \cdots$, where $\gamma \in [0, 1)$ is a discount rate that ensures that the sum remains finite for infinite horizon.

A stationary **policy** of the agent in an MDP is a mapping $\pi(s)$ from states to actions, as in Section 2.4. Clearly, different policies will produce different discounted future rewards, since each policy will take the agent through different trajectories in the state space. The optimal **value** of a state s for the particular agent is defined as the maximum discounted future reward the agent can receive in state s by following some policy:

$$V^*(s) = \max_{\pi} E\left[\sum_{t=0}^{\infty} \gamma^t R(s_t, a_t) | s_0 = s, a_t = \pi(s_t) \right] \qquad (7.1)$$

where the expectation operator $E[\cdot]$ averages over the stochastic transitions. Similarly, the optimal **Q-value** of a state s and action a of the agent is the maximum discounted future reward the agent can receive after taking action a in state s:

$$Q^*(s, a) = \max_{\pi} E\left[\sum_{t=0}^{\infty} \gamma^t R(s_t, a_t) | s_0 = s, a_0 = a, a_{t>0} = \pi(s_t) \right]. \qquad (7.2)$$

A policy $\pi^*(s)$ that achieves the maximum in (7.1) or (7.2) is an optimal policy for the agent. For an MDP there is always an optimal policy that is deterministic and stationary. Deterministic means that $\pi^*(s)$ specifies a single action per state. Stationary means that every time the agent visits a state s, the optimal action to take at s is always $\pi^*(s)$. An optimal policy is greedy with respect to V^* or Q^*, as we have seen in Section 2.4:

$$\pi^*(s) \in \arg\max_a Q^*(s, a). \tag{7.3}$$

Note that there can be many optimal policies in a given task, but they all share a unique V^* and Q^*.

The definition of V^* in (7.1) can be rewritten recursively by making use of the transition model, to get the so-called **Bellman equation**:

$$V^*(s) = \max_a \left[R(s, a) + \gamma \sum_{s'} p(s'|s, a) V^*(s') \right]. \tag{7.4}$$

This is a set of nonlinear equations, one for each state, the solution of which defines the optimal V^*. A similar recursive definition holds for Q-values:

$$Q^*(s, a) = R(s, a) + \gamma \sum_{s'} p(s'|s, a) \max_{a'} Q^*(s', a'). \tag{7.5}$$

7.2.1 Value Iteration

A simple and efficient method for computing optimal values in an MDP when the transition model is available is **value iteration**. In this method we initialize arbitrarily a Q-function (say, with all entries zero), and then we iteratively apply the Bellman equation (7.5) turned into an assignment operation:

$$V(s) := \max_a Q(s, a), \quad \forall s, \tag{7.6}$$

$$Q(s, a) := R(s, a) + \gamma \sum_{s'} p(s'|s, a) V(s'), \quad \forall s, \forall a. \tag{7.7}$$

We repeat the above two equations until V does not change significantly between two consecutive steps. Value iteration converges to the optimal Q^* (and thus to V^* and π^*) for any initialization (Bertsekas, 2001). After we have computed Q^* we can extract an optimal policy π^* using (7.3). As an example, using value iteration in the world of Fig. 2.1 of Chapter 2, with fixed reward $R(s, a) = -1/30$ for each nonterminal state s and action a, and with no discounting, we get the optimal values (utilities) and the optimal policy shown in Fig. 2.2. The reader is encouraged to verify this by implementing the method.

7.2.2 Q-learning

In order to apply the value iteration updates (7.6) and (7.7) we need to know the transition model $p(s'|s, a)$, but in many applications the transition model is unavailable. **Q-learning** is a method for estimating the optimal Q^* (and from that an optimal policy) that does not require knowledge of the transition model. In Q-learning the agent repeatedly interacts with the environment and tries to estimate Q^* by trial-and-error. As in value iteration, the agent initializes a function $Q(s, a)$ for each state–action pair, and then it begins exploring the environment. The exploration generates tuples (s, a, r, s') where s is a state, a is an action taken at state s, $r = R(s, a)$ is a received reward, and s' is a resulting state after executing a. From each such tuple the agent updates its Q-value estimates as

$$V(s') := \max_{a'} Q(s', a') \tag{7.8}$$

$$Q(s, a) := (1 - \lambda) Q(s, a) + \lambda \big[r + \gamma V(s') \big] \tag{7.9}$$

where $\lambda \in (0, 1)$ is a learning rate that regulates convergence.

If all state–action pairs are visited infinitely often and λ decreases slowly with time, Q-learning converges to the optimal Q^* (Watkins and Dayan, 1992). Moreover, this holds irrespectively of the particular exploration policy of the agent. A common exploration policy is the so-called ϵ-**greedy** policy by which in state s the agent selects a random action with probability ϵ, and action $a = \arg\max_{a'} Q(s, a')$ with probability $1 - \epsilon$, where $\epsilon < 1$ is a small number. Alternatively, the agent can choose exploration action a in state s according to a **Boltzmann** distribution

$$p(a|s) = \frac{\exp(Q(s, a)/\tau)}{\sum_{a'} \exp(Q(s, a')/\tau)} \tag{7.10}$$

where τ controls the smoothness of the distribution (and thus the randomness of the choice), and is decreasing with time.

7.3   MARKOV GAMES

In this section we will describe a model for the simultaneous sequential decision making of multiple agents, and describe how reinforcement learning techniques can be extended to cover the case of concurrently learning agents. We will throughout assume that every agent fully observes the current state, as in Chapter 4.

A **Markov game**, also known as **stochastic game**, can be regarded as the multiagent extension of a Markov decision process. It can also be viewed as a collection of coupled strategic games, one per state. Formally, a Markov game is defined by the following primitives:

- Discrete time $t = 0, 1, 2, \ldots$.
- A set of $n > 1$ agents.
- A discrete set of states $s \in S$.
- For each agent i, a discrete set of actions $a_i \in A_i$.
- A stochastic transition model $p(s'|s, a)$ that is conditioned on the *joint* action $a = (a_i)$ at state s.

- For each agent i, a reward function $R_i : S \times A \mapsto \mathbb{R}$, that gives agent i reward $R_i(s, a)$ when *joint* action a is taken at state s.

- A planning horizon, which can be infinite.

Note that the individual reward functions $R_i(s, a)$ define a set of strategic games, one for each state s. A Markov game differs to an MDP in that a transition depends on the joint action of the agents, and that each agent may receive different reward as a result of a joint action. As in MDPs, a policy of an agent i is a mapping $\pi_i(s)$ from states to individual actions. As in strategic games, a joint policy $\pi^* = (\pi_i^*)$ is a Nash equilibrium if no agent has an incentive to unilaterally change its policy; that is, no agent i would like to take at state s an action $a_i \neq \pi_i^*(s)$ assuming that all other agents stick with their equilibrium policies $\pi_{-i}^*(s)$. Contrary to MDPs, in a Markov game an optimal policy of an agent need not be deterministic; we can see this by noticing that a single-state Markov game is just a strategic game, for which we know that deterministic equilibria may not always exist (see Fig. 3.2(a)).

7.3.1 Independent Learning

Learning in a Markov game can be done by each agent separately, ignoring the presence of the other agents in the system. That is, each agent can treat the other agents as part of its environment, without trying to model them or predict their actions. For instance, an agent can use Q-learning to learn its policy, as in standard MDPs, hoping that the resulting policy will perform well. However this approach is inherently flawed: the convergence of Q-learning relies on an underlying transition model that is stationary (does not change with time). This may not be the case with concurrently learning agents, because the (hypothetical) transition model $p(s'|s, a_i)$ of agent i may be changing continuously by the policy of the other agents who are also learning.

Although independent Q-learning cannot be justified theoretically, the method has been employed in practice with reported success (Matarić, 1994, Sen et al., 1994, Tan, 1993). Claus and Boutilier (1998) examine the conditions under which independent Q-learning leads to individual policies that form a Nash equilibrium in a single state coordination problem, arguing that under general conditions the method converges. However, the resulting equilibrium may not be Pareto optimal. Similarly, Wolpert et al. (1999) show that for a constrained class of problems independent Q-learning may converge to a Nash equilibrium.

7.3.2 Coupled Learning

Better results can be obtained if the agents attempt to model each other, in which case their learning algorithms are coupled. A standard approach is to have each agent i maintain an individual value function $V_i(s)$ and an individual Q-function $Q_i(s, a)$, where the latter is defined over joint actions a. In this case, standard value iteration generalizes to Markov games as follows:

$$V_i(s) := C(Q_1(s, a), \ldots, Q_n(s, a)), \quad \forall s \qquad (7.11)$$

$$Q_i(s, a) := R_i(s, a) + \gamma \sum_{s'} p(s'|s, a)V_i(s'), \quad \forall s, \forall a \qquad (7.12)$$

where C is a function that applies some solution concept to the strategic game formed by the $Q_1, \ldots, Q_n$. When the transition model is not available, a corresponding coupled Q-learning update scheme can be derived in which (7.12) is replaced by (7.9), one per agent. Note that such a multiagent Q-learning scheme requires that each agent observes the selected joint action in each step.

Depending on the type of game (zero-sum, general-sum, or coordination game), the function C may compute a Nash equilibrium (Hu and Wellman, 2004, Littman, 1994, 2001), a correlated equilibrium[1] (Greenwald and Hall, 2003), or a coordinated joint action (Kok and Vlassis, 2006). Although successful in practice, the method may not always be able to compute an optimal equilibrium policy (Zinkevich et al., 2006).

7.3.3   Sparse Cooperative Q-learning

Here we address the special case of collaborative agents. When all agents in a team share the same reward function, that is, $R_i(s, a) = R(s, a)$ for each i, the multiagent system can be transformed into one 'big' agent and solved with standard MDP techniques (Lauer and Riedmiller, 2000, Brafman and Tennenholtz, 2003). However, such an approach does not scale very well when the number of agents is large, since the joint action space scales exponentially in the number of agents.

A more general case involves collaborative agents with different individual reward functions. A **collaborative multiagent Markov decision process** (CM-MDP) (Guestrin, 2003) is a Markov game model where each agent cares about maximizing the discounted future *global reward*, where the latter is defined as

$$R(s, a) = \sum_{i=1}^{n} R_i(s, a). \qquad (7.13)$$

Such a model allows for a decentralized learning algorithm where all update steps are local, and intermediate results are communicated by solving a global coordination game. In particular, following the general multiagent learning approach of Section 7.3.2, we can derive a multiagent

---

[1]A correlated equilibrium is a generalization of a mixed-strategy Nash equilibrium where the mixed strategies of the agents can be correlated.

Q-learning algorithm for a CM-MDP. Assuming that each agent i observes tuple (s, a, r_i, s') with $r_i = R_i(s, a)$, we obtain:

$$V_i(s') := Q_i(s', a^*), \quad \text{where} \quad a^* \in \arg\max_{a'} \sum_{i=1}^{n} Q_i(s', a'), \qquad (7.14)$$

$$Q_i(s, a) := (1 - \lambda) Q_i(s, a) + \lambda [r_i + \gamma V_i(s')]. \qquad (7.15)$$

Note that the Q-learning update rule (7.15) is fully decentralized (each agent applies a local update step separately), while (7.14) involves computing a coordinated joint action (a Pareto optimal Nash equilibrium) in a global coordination game with common payoffs $Q(s, a) = \sum_{i=1}^{n} Q_i(s, a)$. The latter can be carried out with a coordination algorithm as in Chapter 4.

When the optimal global Q-function of the task is decomposable as $Q^*(s, a) = \sum_{i=1}^{n} Q_i^*(s, a)$, we can easily show (by summing (7.15) over n) that the above Q-learning algorithm converges to an optimal joint policy. Such a decomposition of Q^* does not hold in general, but there are cases where Q^* may indeed by decomposable (Wiegerinck et al., 2006). In these cases, and assuming a properly decreasing learning rate λ, the above Q-learning algorithm will be optimal.

If each local Q-function $Q_i(s, a)$ is stored as a table (one entry for each state and joint action), the above approach will not scale to many agents. Alternatively we can use a coordination graph approach to represent the global Q-function (see Section 4.4). For instance, we can assume a decomposition $Q(s, a) = \sum_{i=1}^{n} Q_i(s, a)$, where now each local term Q_i may depend on few actions (say, of the neighbors of i in the graph). When such a sparse decomposition of the Q-function is assumed, the coordination step in (7.14) can be carried out (exactly or approximately) by the techniques presented in Section 4.4. In this case, the multiagent Q-learning algorithm (7.14)–(7.15) has been dubbed *Sparse Cooperative Q-learning* (Kok and Vlassis, 2006). Various representations of the local Q-functions can be used, for instance a representation using a functions approximator (Guestrin et al., 2002b). A related approach that uses different local update rules has been proposed by Schneider et al. (1999).

7.4   THE PROBLEM OF EXPLORATION

A critical issue in reinforcement learning is how the agent(s) should explore an unknown environment. An optimal exploration policy would be one that accumulates much reward as fast as possible, while at the same time explores the environment as thoroughly as possible. This is known to be a difficult problem (Kakade, 2003). One approach that is becoming popular recently, termed **Bayesian reinforcement learning**, is based on the idea that optimal exploration can be cast as optimal planning in the space of beliefs over the unknown model parameters, assuming a prior belief over those parameters. Planning in such a space is however intractable, and except for the simplest cases, approximations are needed. Poupart et al. (2006) derive the optimal exploration policy for the standard case of discrete MDPs, but the approach may not scale in large problems (we refer to that paper for more details and references). Chalkiadakis and Boutilier (2003) describe a Bayesian reinforcement learning approach in a problem involving a team of collaborative agents.

In a multiagent system, if we abandon the idea of collecting reward fast, and only care about asymptotic guarantees (like convergence of Q-learning in the limit), we can often simulate in the multiagent system the conditions that ensure asymptotic convergence of single-agent learning algorithms. For instance, in Sparse Cooperative Q-learning, the team can choose a joint action at state s by sampling from a Boltzmann distribution over joint actions using the current $Q(s, a)$. When communication is not available, this can be achieved by having all agents use the same random number generator (and same seed).

Alternatively, the following exploration strategy can be used. At time t and state s, agent i chooses at random k joint actions from the last m $(m > k)$ observed joint actions $\{a^{t-m}, \ldots, a^{t-1}\}$ taken by the agents in state s. Then each agent i computes the relative frequency of each a_{-i}, that is, how many times out of k each a_{-i} was taken by the other agents at state s. This results in an empirical distribution $\mu_i(s, a_{-i})$ that reflects the belief of agent i over the joint action a_{-i} of all other agents at state s. Using such an empirical belief, agent i can now compute an expected payoff $u_i(s, a_i)$ for action a_i at state s as

$$u_i(s, a_i) = \sum_{a_{-i}} \mu_i(s, a_{-i}) Q(s, [a_i, a_{-i}]) \qquad (7.16)$$

and then randomly choose a best-response action $a_i^* \in \arg\max_{a_i} u_i(s, a_i)$.

Variations of this approach have been used by Peyton Young (1993), Claus and Boutilier (1998), and Wang and Sandholm (2003). The approach bears resemblance to **fictitious play**, a learning method in games in which a number of agents interact repeatedly with each other, aiming at reaching an equilibrium (Fudenberg and Levine, 1998).

7.5   NOTES AND FURTHER READING

Single-agent reinforcement learning is treated in the books of Bertsekas and Tsitsiklis (1996) and Sutton and Barto (1998). Fudenberg and Levine (1998), Young (2004), and Greenwald (2007) cover the subject of (reinforcement) learning in games. In the last couple of years several multiagent learning algorithms have appeared in the literature. Besides the works cited in this chapter, notable recent works include those by Bagnell and Ng (2006), Bowling (2005), Brafman and Tennenholtz (2004), Conitzer and Sandholm (2003),

Lagoudakis and Parr (2003), Moallemi and Van Roy (2004), Peshkin et al. (2000), Singh et al. (2000), Tesauro (2004), Zinkevich et al. (2006). **Multiagent reinforcement learning is still a young field;** Shoham et al. (2007) **and** Gordon (2007) **identify several research agendas that can be used for guiding research and evaluating progress.**

Bibliography

Adamatzky, A. and Komosinski, M., editors (2005). *Artificial Life Models in Software*. Springer-Verlag, Berlin.

Arnborg, S., Corneil, D. G., and Proskurowski, A. (1987). Complexity of finding embeddings in a k-tree. *SIAM Journal on Algebraic and Discrete Methods*, 8(2):277–284.

Aumann, R. J. (1976). Agreeing to disagree. *Annals. of Statistics*, 4(6):1236–1239.

Bagnell, J. A. and Ng, A. Y. (2006). On local rewards and the scalability of distributed reinforcement learning. In Weiss, Y., Schölkopf, B., and Platt, J., editors, *Advances in Neural Information Processing Systems 18*. MIT Press, Cambridge, MA.

Bellman, R. (1961). *Adaptive Control Processes: a Guided Tour*. Princeton University Press.

Bernstein, D. S., Givan, R., Immerman, N., and Zilberstein, S. (2002). The complexity of decentralized control of Markov decision processes. *Mathematics of Operations Research*, 27(4):819–840. doi:10.1287/moor.27.4.819.297

Bertsekas, D. P. (2001). *Dynamic Programming and Optimal Control*, volume I and II. Athena Scientific, 2nd edition.

Bertsekas, D. P. and Tsitsiklis, J. N. (1996). *Neuro-dynamic Programming*. Athena Scientific.

Bordini, R., Dastani, M., Dix, J., and El Fallah Seghrouchni, A., editors (2005). *Multi-Agent Programming: Languages, Platforms and Applications*. Springer, Berlin.

Boutilier, C. (1996). Planning, learning and coordination in multiagent decision processes. In *Proc. Conf. on Theoretical Aspects of Rationality and Knowledge*, Renesse, The Netherlands.

Bowling, M. (2005). Convergence and no-regret in multiagent learning. In Saul, L. K., Weiss, Y., and Bottou, L., editors, *Advances in Neural Information Processing Systems 17*, pp. 209–216. MIT Press, Cambridge, MA.

Brafman, R. I. and Tennenholtz, M. (2003). Learning to coordinate efficiently: A model based approach. *Journal of Artificial Intelligence Research*, 19:11–23.

Brafman, R. I. and Tennenholtz, M. (2004). Efficient learning equilibrium. *Artificial Intelligence*, 159(1-2):27–47. doi:10.1016/j.artint.2004.04.013

Castelpietra, C., Iocchi, L., Nardi, D., Piaggio, M., Scalzo, A., and Sgorbissa, A. (2000). Coordination among heterogenous robotic soccer players. In *Proc. IEEE/RSJ Int. Conf. on Intelligent Robots and Systems*, Takamatsu, Japan.

Chalkiadakis, G. and Boutilier, C. (2003). Coordination in multiagent reinforcement learning: A Bayesian approach. In *Proc. 2nd Int. Joint Conf. on Autonomous Agents and Multiagent Systems*, Melbourne, Australia.

Clarke, E. H. (1971). Multipart pricing of public goods. *Public Choice*, 11:17–33. doi:10.1007/BF01726210

Claus, C. and Boutilier, C. (1998). The dynamics of reinforcement learning in cooperative multiagent systems. In *Proc. 15th Nation. Conf. on Artificial Intelligence*, Madison, WI.

Cohen, P. R. and Levesque, H. J. (1991). Teamwork. *Nous*, 25(4):487–512.

Conitzer, V. (2006). *Computational Aspects of Preference Aggregation*. PhD thesis, Computer Science Department, Carnegie Mellon University.

Conitzer, V. and Sandholm, T. (2003). AWESOME: A general multiagent learning algorithm that converges in self-play and learns a best response against stationary opponents. In *Proc. 20th Int. Conf. on Machine Learning*, Washington, DC, USA.

Conitzer, V. and Sandholm, T. (2007). Incremental mechanism design. In *Proc. 20th Int. Joint Conf. on Artificial Intelligence*, Hyderabad, India.

Conte, R. and Dellarocas, C., editors (2001). *Social Order in Multiagent Systems*. Kluwer Academic, Dordrecht.

Cramton, P., Shoham, Y., and Steinberg, R., editors (2006). *Combinatorial Auctions*. MIT Press, Cambridge, MA.

Dechter, R. (2003). *Constraint Processing*. Morgan Kaufmann, San Mates, CA.

Decker, K. and Lesser, V. R. (1995). Designing a family of coordination algorithms. In *Proc. 1st Int. Conf. on Multi-Agent Systems*, San Francisco, CA.

Durfee, E. H. and Lesser, V. R. (1987). Using partial global plans to coordinate distributed problem solvers. In *Proc. 10th Int. Joint Conf. on Artificial Intelligence*, Milan, Italy.

Emery-Montemerlo, R., Gordon, G., Schneider, J., and Thrun, S. (2005). Game theoretic control for robot teams. In *Proc. IEEE Int. Conf. on Robotics and Automation*, Barcelona, Spain.

Fagin, R., Halpern, J., Moses, Y., and Vardi, M. (1995). *Reasoning about Knowledge*. The MIT Press, Cambridge, MA.

Ferber, J. (1999). *Multi-Agent Systems: An Introduction to Distributed Artificial Intelligence*. Addison-Wesley, Reading, MA.

Fudenberg, D. and Levine, D. K. (1998). *The Theory of Learning in Games*. MIT Press, Cambridge, MA.

Geanakoplos, J. (1992). Common knowledge. *Journal of Economic Perspectives*, 6(4):53–82.

Gibbard, A. (1973). Manipulation of voting schemes: a general result. *Econometrica*, 41:587–601. doi:10.2307/1914083

Gibbons, R. (1992). *Game Theory for Applied Economists*. Princeton University Press, Princeton, NJ.

Gilbert, N. and Doran, J., editors (1994). *Simulating Societies: The Computer Simulation of Social Phenomena*. UCL Press, London.

Gmytrasiewicz, P. J. and Durfee, E. H. (2001). Rational communication in multi-agent environments. *Autonomous Agents and Multi-Agent Systems*, 4:233–272. doi:10.1023/A:1011495811107

Gordon, G. J. (2007). Agendas for multi-agent learning. *Artificial Intelligence*. doi: 10.1016/j.artint.2006.12.006.

Greenwald, A. (2007). *The Search for Equilibrium in Markov Games (Synthesis Lectures on Artificial Intelligence and Machine Learning)*. Morgan & Claypool Publishers, San Rafael, CA.

Greenwald, A. and Hall, K. (2003). Correlated-Q learning. In *Proc. 20th Int. Conf. on Machine Learning*, Washington, DC, USA.

Groves, T. (1973). Incentives in teams. *Econometrica*, 41:617–631. doi:10.2307/1914085

Guestrin, C. (2003). *Planning Under Uncertainty in Complex Structured Environments*. PhD thesis, Computer Science Department, Stanford University.

Guestrin, C., Koller, D., and Parr, R. (2002a). Multiagent planning with factored MDPs. In Dietterich, T. G., Becker, S., and Ghahramani, Z., editors, *Advances in Neural Information Processing Systems 14*. The MIT Press, Cambridge, MA.

Guestrin, C., Lagoudakis, M., and Parr, R. (2002b). Coordinated reinforcement learning. In *Proc. 19th Int. Conf. on Machine Learning*, Sydney, Australia.

Hansen, E. A., Bernstein, D. S., and Zilberstein, S. (2004). Dynamic programming for partially observable stochastic games. In *Proc. 19th National Conf. on Artificial Intelligence*, San Jose, CA.

Harsanyi, J. C. (1967). Games with incomplete information played by 'Bayesian' players, Parts I, II, and III. *Management Science*, 14:159–182, 320–334, and 486–502.

Harsanyi, J. C. and Selten, R. (1988). *A General Theory of Equilibrium Selection in Games*, MIT Press, Cambridge, MA.

Hu, J. and Wellman, M. P. (2004). Nash Q-learning for general-sum stochastic games. *Journal of Machine Learning Research*, 4:1039–1069. doi:10.1162/1532443041827880

Huhns, M. N., editor (1987). *Distributed Artificial Intelligence*. Pitman, Morgan Kaufmann, San Mateo, CA.

Jennings, N. R. (1996). Coordination techniques for distributed artificial intelligence. In O'Hare, G. M. P. and Jennings, N. R., editors, *Foundations of Distributed Artificial Intelligence*, pp. 187–210. John Wiley & Sons, New York.

Kakade, S. (2003). *On the Sample Complexity of Reinforcement Learning*. PhD thesis, Gatsby Computational Neuroscience Unit, University College London.

Kitano, H., Tambe, M., Stone, P., Veloso, M., Coradeschi, S., Osawa, E., Matsubara, H., Noda, I., and Asada, M. (1997). The RoboCup synthetic agent challenge 97. In *Proc. Int. Joint Conf. on Artificial Intelligence*, Nagoya, Japan.

Kok, J. R., Spaan, M. T. J., and Vlassis, N. (2005). Non-communicative multi-robot coordination in dynamic environments. *Robotics and Autonomous Systems*, 50(2–3):99–114. doi:10.1016/j.robot.2004.08.003

Kok, J. R. and Vlassis, N. (2006). Collaborative multiagent reinforcement learning by payoff propagation. *Journal of Machine Learning Research*, 7:1789–1828.

Kowalczyk, W. and Vlassis, N. (2005). Newscast EM. In Saul, L. K., Weiss, Y., and Bottou, L., editors, *Advances in Neural Information Processing Systems 17*, pp. 713–720. MIT Press, Cambridge, MA.

Lagoudakis, M. G. and Parr, R. (2003). Learning in zero-sum team Markov games using factored value functions. In Becker, S., Thrun, S., and Obermayer, K., editors, *Advances in Neural Information Processing Systems 15*, MIT Press, Cambridge, MA.

Lauer, M. and Riedmiller, M. (2000). An algorithm for distributed reinforcement learning in cooperative multi-agent systems. In *Proc. 17th Int. Conf. on Machine Learning*, Stanford University, USA.

Lesser, V., Ortiz Jr., C. L., and Tambe, M., editors (2003). *Distributed Sensor Networks: A Multiagent Perspective*. Kluwer Academic, Dodrecht.

Lewis, D. K. (1969). *Conventions: A Philosophical Study*. Harvard University Press, Cambridge.

Littman, M. L. (1994). Markov games as a framework for multi-agent reinforcement learning. In *Proc. 11th Int. Conf. on Machine Learning*, San Francisco, CA.

Littman, M. L. (2001). Friend-or-foe Q-learning in general-sum games. In *Proc. 18th Int. Conf. on Machine Learning*, San Francisco, CA.

Matarić, M. J. (1994). Reward functions for accelerated learning. In *Proc. 11th Int. Conf. on Machine Learning*, San Francisco, CA.

Moallemi, C. C. and Van Roy, B. (2004). Distributed optimization in adaptive networks. In Thrun, S., Saul, L., and Schölkopf, B., editors, *Advances in Neural Information Processing Systems 16*. MIT Press, Cambridge, MA.

Modi, P. J., Shen, W.-M., Tambe, M., and Yokoo, M. (2005). ADOPT: Asynchronous distributed constraint optimization with quality guarantees. *Artificial Intelligence*, 161 (1–2):149–180. doi:10.1016/j.artint.2004.09.003

Nash, J. F. (1950). Equilibrium points in n-person games. *Proceedings of the National Academy of Sciences*, 36:48–49. doi:10.1073/pnas.36.1.48

Nisan, N. (1999). Algorithms for selfish agents. In *Proc. 16th Symp. on Theoret. Aspects of Computer Science*, Trier, Germany.

Nisan, N., Tardos, E., and Vazirani, V., editors (2007). *Algorithmic Game Theory*. Cambridge University Press, Cambridge.

Noriega, P. and Sierra, C., editors (1999). *Agent Mediated Electronic Commerce*. Lecture Notes in Artificial Intelligence 1571. Springer, Berlin.

O'Hare, G. M. P. and Jennings, N. R., editors (1996). *Foundations of Distributed Artificial Intelligence*. John Wiley & Sons, New York.

Oliehoek, F. A. and Vlassis, N. (2007). Q-value functions for decentralized POMDPs. In *Proc. of Int. Joint Conf. on Autonomous Agents and Multi Agent Systems*, Honolulu, Hawai'i.

Osborne, M. J. (2003). *An Introduction to Game Theory*. Oxford University Press, Oxford.

Osborne, M. J. and Rubinstein, A. (1994). *A Course in Game Theory*. MIT Press, Cambridge, MA.

Papadimitriou, C. H. (2001). Algorithms, games, and the Internet. In *Proc. 33rd Ann. ACM Symp. on Theory of Computing*, Heraklion, Greece.

Papadimitriou, C. H. and Tsitsiklis, J. N. (1987). The complexity of Markov decision processes. *Mathematics of Operations Research*, 12(3):441–450.

Parkes, D. C. (2001). *Iterative Combinatorial Auctions: Achieving Economic and Computational Efficiency*. PhD thesis, Computer and Information Science, University of Pennsylvania.

Parkes, D. C. and Shneidman, J. (2004). Distributed implementations of Vickrey-Clarke-Groves mechanisms. In *Proc. 3nd Int. Joint Conf. on Autonomous Agents and Multiagent Systems*, New York, USA.

Paskin, M. A., Guestrin, C. E., and McFadden, J. (2005). A robust architecture for distributed inference in sensor networks. In *Proc. 4th Int. Symp. on Information Processing in Sensor Networks*, Los Angeles, CA.

Pearce, J. P. and Tambe, M. (2007). Quality guarantees on k-optimal solutions for distributed constraint optimization problems. In *Proc. 20th Int. Joint Conf. on Artificial Intelligence*, Hyderabad, India.

Pearl, J. (1988). *Probabilistic Reasoning in Intelligent Systems*. Morgan Kaufman, San Mateo, CA.

Peshkin, L., Kee-Eung, K., Meuleau, N., and Kaelbling, L. (2000). Learning to cooperate via policy search. In *Proc. 16th Int. Conf. on Uncertainty in Artificial Intelligence*, San Francisco, CA.

Peyton Young, H. (1993). The evolution of conventions. *Econometrica*, 61(1):57–84. doi:10.2307/2951778

Porter, R., Nudelman, E., and Shoham, Y. (2004). Simple search methods for finding a Nash equilibrium. In *Proc. 19th National Conf. on Artificial Intelligence*, San Jose, CA.

Poupart, P., Vlassis, N., Hoey, J., and Regan, K. (2006). An analytic solution to discrete Bayesian reinforcement learning. In *Proc. Int. Conf. on Machine Learning*, Pittsburgh, USA.

Roumeliotis, S. I. and Bekey, G. A. (2002). Distributed multi-robot localization. *IEEE Trans. Robotics and Automation*, 18(5):781–795. doi:10.1109/TRA.2002.803461

Russell, S. J. and Norvig, P. (2003). *Artificial Intelligence: a Modern Approach*. Prentice Hall, Englewood Cliffs, NJ, 2nd edition.

Samet, D. (2006). Agreeing to disagree: the non-probabilistic case. Unpublished manuscript (available from http://www.tau.ac.il/~samet).

Sandell, N. R., Varayia, P., Athans, M., and Safonov, M. G. (1978). Survey of decentralized control methods for large-scale systems. *IEEE Trans. Automatic Control*, AC-23(2):108–128. doi:10.1109/TAC.1978.1101704

Sandholm, T. (1999). Distributed rational decision making. In Weiss, G., editor, *Multiagent Systems: A Modern Introduction to Distributed Artificial Intelligence*, pp. 201–258. MIT Press, Cambridge, MA.

Schneider, J., Wong, W.-K., Moore, A., and Riedmiller, M. (1999). Distributed value functions. In *Proc. Int. Conf. on Machine Learning*, Bled, Slovenia.

Sen, S., Sekaran, M., and Hale, J. (1994). Learning to coordinate without sharing information. In *Proc. 12th Nation. Conf. on Artificial Intelligence*, Seattle, WA.

Shoham, Y. and Leyton-Brown, K. (2007). *Multi-Agent Systems*. Cambridge University Press, Cambridge.

Shoham, Y., Powers, R., and Grenager, T. (2007). If multi-agent learning is the answer, what is the question? *Artificial Intelligence*. doi: 10.1016/j.artint.2006.02.006.

Shoham, Y. and Tennenholtz, M. (1992). On the synthesis of useful social laws for artificial agent societies. In *Proc. 10th Nation. Conf. on Artificial Intelligence*, San Diego, CA.

Singh, M. P. (1994). *Multiagent Systems: A Theoretical Framework for Intentions, Know-How, and Communications*. Springer-Verlag, Berlin.

Singh, S., Kearns, M., and Mansour, Y. (2000). Nash convergence of gradient dynamics in general-sum games. In *Proc. 16th Int. Conf. on Uncertainty in Artificial Intelligence*, San Francisco, CA.

Smith, R. G. (1980). The contract net protocol: high-level communication and control in a distributed problem solver. *IEEE Trans. Computers*, C-29(12):1104–1113.

Spaan, M. T. J. and Vlassis, N. (2005). Perseus: Randomized point-based value iteration for POMDPs. *Journal of Artificial Intelligence Research*, 24:195–220.

Stone, P. (2000). *Layered Learning in Multiagent Systems: A Winning Approach to Robotic Soccer*. MIT Press, Cambridge, MA.

Stone, P. and Veloso, M. (2000). Multiagent systems: a survey from a machine learning perspective. *Autonomous Robots*, 8(3).

Sutton, R. S. and Barto, A. G. (1998). *Reinforcement Learning: An Introduction*. MIT Press, Cambridge, MA.

Sycara, K. (1998). Multiagent systems. *AI Magazine*, 19(2):79–92.

Symeonidis, A. L. and Mitkas, P. A., editors (2006). *Agent Intelligence Through Data Mining*. Springer, Berlin.

Tan, M. (1993). Multi-agent reinforcement learning: Independent vs. cooperative agents. In *Proc. 10th Int. Conf. on Machine Learning*, Amherst, MA.

Terzopoulos, D. (1999). Artificial life for computer graphics. *Commun. ACM*, 42(8):32–42. doi:10.1145/310930.310966

Tesauro, G. (2004). Extending Q-learning to general adaptive multi-agent systems. In Thrun, S., Saul, L., and Schölkopf, B., editors, *Advances in Neural Information Processing Systems 16*. MIT Press, Cambridge, MA.

Thorndike, E. L. (1898). *Animal Intelligence: An Experimental Study of the Associative Processes in Animals*. PhD thesis, Columbia University.

Vickrey, W. (1961). Counterspeculation, auctions, and competitive sealed tenders. *Journal of Finance*, 16:8–37. doi:10.2307/2977633

Vidal, J. M. (2007). *Fundamentals of Multiagent Systems with NetLogo Examples*. Electronically available at www.multiagent.com.

Vlassis, N., Elhorst, R. K., and Kok, J. R. (2004). Anytime algorithms for multiagent decision making using coordination graphs. In *Proc. Int. Conf. on Systems, Man and Cybernetics*, The Hague, The Netherlands.

von Neumann, J. and Morgenstern, O. (1944). *Theory of Games and Economic Behavior*. John Wiley & Sons, New York.

von Stengel, B. (2007). Equilibrium computation for two-player games in strategic and extensive form. In Nisan, N., Tardos, E., and Vazirani, V., editors, *Algorithmic Game Theory*. Cambridge University Press, Cambridge.

Wainwright, M. J., Jaakkola, T. S., and Willsky, A. S. (2004). Tree consistency and bounds on the performance of the max-product algorithm and its generalizations. *Statistics and Computing*, 14:143–166. doi:10.1023/B:STCO.0000021412.33763.d5

Wang, X. and Sandholm, T. (2003). Reinforcement learning to play an optimal Nash equilibrium in team Markov games. In Becker, S., Thrun, S., and Obermayer, K., editors, *Advances in Neural Information Processing Systems 15*, MIT Press, Cambridge, MA.

Watkins, C. J. C. H. and Dayan, P. (1992). Q-learning. *Machine Learning*, 8(3):279–292.

Weiss, G., editor (1999). *Multiagent Systems: a Modern Approach to Distributed Artificial Intelligence*. MIT Press, Cambridge, MA.

Wiegerinck, W., van den Broek, B., and Kappen, B. (2006). Stochastic optimal control in continuous space-time multi-agent systems. In *Proc. 22th Int. Conf. on Uncertainty in Artificial Intelligence*, MIT Press, Cambridge, MA.

Wolpert, D., Wheeler, K., and Tumer, K. (1999). General principles of learning-based multi-agent systems. In *Proc. 3rd Int. Conf. on Autonomous Agents*, Seattle, WA.

Wooldridge, M. (2002). *An Introduction to MultiAgent Systems*. John Wiley & Sons, New York.

Xiang, Y. (2002). *Probabilistic Reasoning in Multiagent Systems: A Graphical Models Approach*. Cambridge University Press, Cambridge.

Yokoo, M. (2000). *Distributed Constraint Satisfaction: Foundation of Cooperation in Multi-agent Systems*. Springer, Berlin.

Young, H. P. (2004). *Strategic learning and its limits*. Oxford University Press, Oxford.

Zhang, W., Wang, G., Xing, Z., and Wittenberg, L. (2005). Distributed stochastic search and distributed breakout: Properties, comparison and applications to constraint optimization problems in sensor networks. *Artificial Intelligence*, 161(1-2):55–87. doi:10.1016/j.artint.2004.10.004

Zinkevich, M., Greenwald, A., and Littman, M. L. (2006). Cyclic equilibria in Markov games. In Weiss, Y., Schölkopf, B., and Platt, J., editors, *Advances in Neural Information Processing Systems 18*, pp. 1641–1648. MIT Press, Cambridge, MA.

Author Biography

Nikos Vlassis was born in 1970 in Corinth, Greece. He received an MSc (1993) and a PhD (1998) in Electrical and Computer Engineering from the National Technical University of Athens, Greece. In 1998 he joined the Informatics Institute of the University of Amsterdam, The Netherlands, as research fellow, and in 1999 he visited the Electrotechnical Laboratory (ETL, currently AIST) in Tsukuba, Japan, with a scholarship from the Japan Industrial Technology Association (MITI). From 2000 until 2006 he held an Assistant Professor position in the Informatics Institute of the University of Amsterdam, The Netherlands. Since 2007 he holds an Assistant Professor position in the Department of Production Engineering and Management of the Technical University of Crete, Greece. He is coauthor of about 100 papers on various topics in the fields of machine learning, multiagent systems, robotics, and computer vision, and has received numerous citations. Awards that he has received include the Dimitris Chorafas Foundation prize for young researchers in Engineering and Technology (Luzern, Switzerland, 1998), best-teacher mention at the University of Amsterdam (2001–2005), best scientific paper award with the paper 'Using the max-plus algorithm for multiagent decision making in coordination graphs' in the annual RoboCup symposium (2005), and various distinctions with the UvA Trilearn robot soccer team including the 1st position at the RoboCup world championship (2003). His current research interests are in the areas of robotics, machine learning, and stochastic optimal control.

Printed in the United States
by Baker & Taylor Publisher Services